# A Modern Introduction to Ancient Indian Mathematics

## Second Edition

T.S.BhanuMurthy

*Former Director*

*RamanujamIn stituteo f*

*AdvancedSt udyi nM athematics*

*University of Madras*

*Chennai*

PUBLISHING FOR ONE WORLD

**NEW AGE INTERNATIONAL (P) LIMITED, PUBLISHERS**

New Delhi • Bangalore • Chennai • Cochin • Guwahati • Hyderabad

Jalandhar • Kolkata • Lucknow • Mumbai • Ranchi

Visit us at **www.newagepublishers.com**

**Published by New Age International (P) Ltd., Publishers**
First Edition: 1992
Second Edition: 2009



***Branches:***

- 36, Malikarjuna Temple Street, Opp. ICWA, Basavanagudi, **Bangalore**. ✆ (080) 26677815
- 26, Damodaran Street, T. Nagar, **Chennai**. ✆ (044) 24353401
- Hemsen Complex, Mohd. Shah Road, Paltan Bazar, Near Starline Hotel, **Guwahati**. ✆ (0361) 2543669
- No. 105, 1st Floor, Madhiray Kaveri Tower, 3-2-19, Azam Jahi Road, Nimboliadda, **Hyderabad**. ✆ (040) 24652456
- RDB Chambers (Formerly Lotus Cinema) 106A, Ist Floor, S.N. Banerjee Road, **Kolkata**. ✆ (033) 22275247
- 18, Madan Mohan Malviya Marg, **Lucknow**. ✆ (0522) 2209578
- 142C, Victor House, Ground Floor, N.M. Joshi Marg, Lower Parel, **Mumbai**. ✆ (022) 24927869
- 22, Golden House, Daryaganj, **New Delhi**. ✆ (011) 23262368, 23262370

**ISBN: 978-81-224-2600-7**

**Rs. 195.00**

**C-08-10-3167**

Printed in India at A. H. Printpack, Delhi.
Typeset at Pagitek Graphics, Delhi

**PUBLISHING FOR ONE WORLD**
**NEW AGE INTERNATIONAL (P) LIMITED, PUBLISHERS**
4835/24, Ansari Road, Daryaganj, New Delhi - 110002
Visit us at **www.newagepublishers.com**

To the memory of my

*revered parents*

**Srimathi Sharadamma**

**&**

**Sri Vedantha Subbiah Sastri**

# PREFACE TO THE SECOND EDITION

There are three supplements that have been added to the book:

- (I) The BKT Division Algorithm which was omitted in the first edition.
- (II) A thoroughly revised and simplified version of Bhaskara's Chakravala Method which follows more faithfully its essence. It is the shift of emphasis from $\{a_\lambda\}$ to $(m_\lambda, h_\lambda\}$.
- (III) The solutions of the equation $f(x, y) = a$, where $a \in \mathbf{Z}$, $f(x, y) \in \mathbf{Z}(x, y)$ have been studied from ancient times in Bharatvarsh, China, Greece etc. It is after Hermite, who proved the transcendence of $\varepsilon$, that the modern development has started; in the works of Thue, Siegel, Gelfond, Schneider, Roth, Baker, Wolfgang Schmidt. The ancient subject has retained its youth, beauty and fascination. The target is of this little book is the student reader whom we wish to entice to this study. An attempt is made in Supplement III to give a few glimpses of recent developments.

**T.S. Bhanu Murthy**

# PREFACE TO THE FIRST EDITION

The project of writing this book was undertaken at the suggestion of Swami Paramananda Bhārati. Initially the purpose was to present proofs of the propositions stated by Swami Bhārati Krishna Tirtha in his well-known treatise VEDIC MATHEMATICS. However it was decided to broaden the scope of the book to serve as a modern introduction to the heritage of mathematical learning in Ancient India. The intention of the book is to invite the attention of teachers of Mathematics in our country to this rich and ancient tradition of mathematics.

The *Śulba Sūtras* in the Vedas is a manual of geometrical constructions; the sūtras cannot be later than 5th century BC and may be as old as 9th century BC. The *Taittiriya Samhitā* of *Yajurveda* gives the measurements of *Mahavèdi* having to do with a right triangle of sides 15 and 26 and hypotenuse 39. The Hypotenuse Theorem: $c^2 = a^2 + b^2$, ($c$ = hypotenuse, $a$ and $b$ are the sides of a right triangle) is one of the important achievements of this early period. This theorem is usually attributed to Pythagoras (540 BC). There are many proofs of this theorem by Euclid, Bhāskara, Leonardo da Vinci, .... Kātyāyana gives a construction of a right triangle with sides $\frac{n^2-1}{2}\cdot a$, $n \cdot a$ and hypotenuse $\frac{n^2+1}{2}a$. Such geometrical constructions were used on building the *Vedis*; these constructions imply a knowledge of the properties of similar triangles. *Āpastamba* and *Kātyāyana* list several right triangles. Similarly, to quote another instance, E.J.H. Mackey [≪ Further Excavations at Mohenjo Daro ≫ (1938)] has said: ≪ It is surprising to find that an instrument was actually used for drawing circles in the Indus Valley as early as 2500 BC. ≫

It is our misfortune that many links of the past are buried in the sands of time. We arrive at the times of Āryabhata I (475 AD), Brahma Gupta (7th century), Bhāskara II (1150 AD), Sangama

Grāma Mādhava and Narāyana Pandita (14th century). I have made an attempt in my book to dispel the common impression that after the 12th century there was utter stagnation in the development of mathematics in our country.

In 1935 and 1938 there appeared the masterly treatise, in two parts, ≪ History of Hindu Mathematics ≫ by B.B. Datta and A.N. Singh. It gave an absorbing narrative of the development of mathematics in our country starting from the Vedic period to the times of Aryabhata (5th century), Brahma Gupta (7th Century), Bhāskara II (12th century). Dr Bibhuti Bhushan Datta and Dr Avadhesh Narain Singh were dedicated scholars of great competence. This inspiring book was one of the several contributions which appeared between late 19th and early 20th centuries, reviving rightful pride in the historicity of our knowledge among our people and to reaffirm an Indian identity.

Sangama Grāma Mādhava, Paramèswara, Nilakantha, the author of Tantra Sangraha, the author of Kriyākramakari and Yukti Bhāsha— all these belong to a period of intensive development of Mathematics and Astronomy in Kérala from the 14th to the 17th century (See Appendix).

I must emphasize the fact that the present book is not devoted to recounting the historical development of Mathematics in our country. For such a study the reader is referred to the classic treatise of B.B. Datta and A.N. Singh, the monograph of C.N. Srinivasiengar and the monograph of T.A. Sarasvati Amma.

The purpose of my book is two-fold: 1° to provide simple proofs of the propositions stated by Sri Bhārati Krishna Tirtha in his book Vedic Mathematics; 2° to provide a modern introduction to some of the theorems of Aryabhata, Brahma Gupta, Bhaskara, and Màdhava.

A preliminary version of the first two chapters of the book was used as course material in the conference cum workshop on Vedic Mathematics (July 13-15, 1990) organised by Abhinava Vidya Bharati, Bangalore. I would therefore, particularly like to thank Mr. Kirit Joshi of the Rashtriya Veda Vidya Pratishthan for sponsoring the above conference, which has resulted in this book.

I would also like to thank my brother Professor T.S. Nanjundiah for having read a preliminary version of the manuscript and suggesting many improvements.

I express my profound gratitude to Swami Paramānanda Bhārati for the tremendous support that he has given to me throughout the time that the book was being written.

I wish to record my deep appreciation to Sri Bal Menon, Laser Words, Madras for excellent photo typesetting of the text. The preparation of the book would have been impossible without the constant assistance and co-ordination of Sri V. Ramnarayan and Sri S. Chandrasekhar.

I also wish to place on record my sincere thanks to Mr. A.N. Haksar, Chairman Emeritus, ITC Ltd., for the help and guidance given by him.

Lastly, I would like to express my profound gratitude to ITC Bhadrachalam Paperboards Ltd. for sponsoring the book.

**T.S. Bhanu Murthy**

# CONTENTS

CHAPTER I

# INTEGERS

## § 1. THE DECIMAL PLACE VALUE SYSTEM

We denote by **N** the set of all positive integers including zero:

**(N)** $0, 1, 2, 3, \ldots n, \ldots$

and often refer to it as the set of *natural numbers*; and by **Z** the set of all integers both positive and negative and including zero:

**(Z)** $0, \pm 1, \pm 2, \ldots, \pm n, \ldots$

The symbol 0 and the decimal place value notation were invented by ancient Hindu mathematicians. In this system every positive integer is represented as sums of powers of 10: for example,

$$1234 = 4 + 3.10 + 2.10^2 + 1.10^3$$

In general, given a positive integer $A$ we express it uniquely in the form

(1) $$A = a_0 + a_1.10 + a_2.10^2 + \ldots + a_n.10^n$$
$$0 \leq a_k \leq 9, a_n \neq 0, \quad (k = 0, 1, \ldots, n)$$

Here $a_0, a_1, \ldots, a_n$ are called the *digits* of A (in the scale of 10). We write

$$A = a_n a_{n-1} \cdots a_2 a_1 a_0$$

provided there is no confusion with the product of $a_0, a_1, \ldots, a_n$. When we need to avoid this confusion, and wish to emphasise that it is not the product we use the notation

(2) $$A = \langle a_n, a_{n-1}, \ldots, a_1, a_0 \rangle_{10}$$

The suffix 10 herein is to remind us that it is the scale of 10.

Suppose we fix some positive integer $T > 1$, instead of 10, as our scale: $T$ is called the *radix* of our proposed scale.

**Theorem 1 (Euclidean Algorithm)**

*Given a positive integer $m$, every integer $a$ can be* **uniquely** *expressed in the form*

$$a = q.m + r, \qquad 0 \leq r < m$$

Herein $q$ is called the quotient and $r$ the remainder of a when divided by $m$.

**Proof**

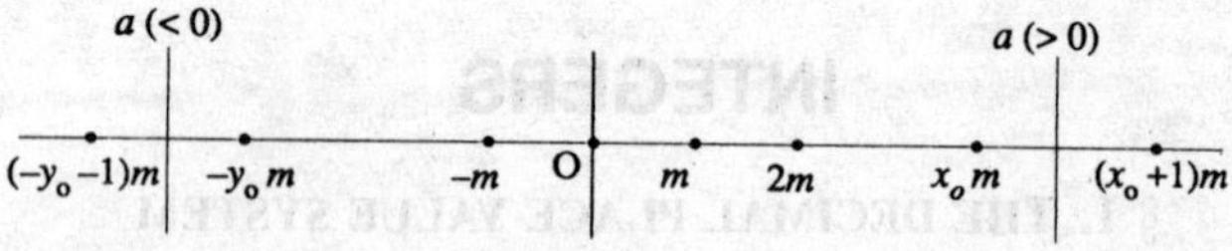


If $a > 0$, we consider the increasing sequence of multiples of m:

$$0, m, 2m, \ldots$$

Let $x_0 m$ be the largest multiple of $m$ that does not exceed $a$, i.e.

$$x_0 m \leq a < (x_0 + 1)m$$

We put $q = x_0$ and have $0 \leq a - qm < m$.
If we take $r = a - qm$ we find

$$a = qm + r, 0 \leq r < m$$

If $a < 0$, we consider the decreasing sequence of multiples of $m$

$$0, -m, -2m, \ldots$$

Determine $y_0 > 0$ such that

$$(-y_0 - 1)m \leq a < -y_0 m$$

and take $q = -y_0 - 1$; then $qm \leq a < (q+1)m$

$$0 \leq a - qm < m$$

$$a = qm + r, \quad 0 \leq r < m \blacksquare$$

We shall use Theorem 1 to express a given positive integer $A$ in the scale of $T$: divide $A$ by $T$, let $q_1$, be the quotient and $a_0$ the remainder, thus,

$$A = a_0 + q_1 T, \quad q_1 < A, 0 \leq a_0 < T$$

Next divide $q_1$ by $T$, let $q_2$ be the quotient and $a_1$ the remainder,

$$q_1 = a_1 + q_2 T, \qquad q_2 < q_1, \quad 0 \leq a_1 < T$$

In this manner we obtain a decreasing sequence of positive integers

$$A > q_1 > q_2 > \ldots > q_n$$

such that

$$
\begin{aligned}
A &= a_0 + q_1 T \\
q_1 &= a_1 + q_2 T \\
q_2 &= a_2 + q_3 T \qquad\qquad 0 \le a_k < T \quad (3) \\
&\vdots \qquad\qquad\qquad\qquad\qquad q_n = a_n \ne 0 \\
q_{n-1} &= a_{n-1} + q_n T \\
q_n &= a_n
\end{aligned}
$$

The process terminates as soon as we reach $q_n$ with $q_n < T$. Multiply the equations (3), in order, by

$$1, T, T^2, \ldots, T^n$$

and add; one gets

$$A = a_0 + a_1 T + a_2 T^2 + \ldots + a_n T^n \tag{4}$$

Since in the division algorithm the $a_k$ are uniquely determined the expansion (4) of $A$ in the scale of T is unique.

**Example 1.** Express 54321 in the scale of 7.

A = 54321 T = 7

$A = a_0 + q_1 T,\quad a_0 = 1,\ q_1 = 7760$

$q_1 = a_1 + q_2 T,\quad a_1 = 4\quad q_2 = 1108$

$q_2 = a_2 + q_3 T,\quad a_2 = 2\quad q_3 = 158$

$q_3 = a_3 + q_4 T,\quad a_3 = 4\quad q_4 = 22$

$q_4 = a_4 + q_5 T,\quad a_4 = 1\quad q_5 = 3\quad (q_5 < T)$

$54321 = 1 + 4.7 + 2.7^2 + 4.7^3 + 1.7^4$

i.e. $54321 = \langle 1, 4, 2, 4, 1 \rangle_7$

$$
\begin{array}{r|l|l}
7 & 54321 & 7760 \\
 & 49 & \\
 & \overline{\ \ 53} & \\
 & \ \ 49 & \\
 & \overline{\ \ \ 42} & \\
 & \ \ \ 42 & \\
 & \overline{\ \ \ \ \ 1} &
\end{array}
$$

7760 = 7.1108 + 4

1108 = 7.158 + 2

158 = 7.22 + 4

22 = 7.3 + 1

*Remark:* The decimal expansion

$$A = 5.2468$$

means that

$$A = 5 + \frac{2}{10} + \frac{4}{10^2} + \frac{6}{10^3} + \frac{8}{10^4}$$

Similarly

$$0.001234 = \frac{1}{10^3} + \frac{2}{10^4} + \frac{3}{10^5} + \frac{4}{10^6}$$

*Definition 1.* Let $\alpha$ be a real number. We denote by $[\alpha]$ the greatest integer not exceeding $\alpha$, thus

(5) $$n = [\alpha], \quad n \leq \alpha < n+1$$

We put $\{\alpha\} = \alpha - [\alpha] = \alpha - n$; since $0 \leq \alpha - n < 1$ we have

(6) $$\alpha = [\alpha] + \{\alpha\}, \; 0 \leq \{\alpha\} < 1, \quad [\alpha] \text{ integer}$$

We call $[\alpha]$ the integral part of $\alpha$ and $\{\alpha\}$ the fractional part of $\alpha$

Since $n \leq \alpha < n+1$ can also be written as

$$\alpha - 1 < n \leq \alpha$$

we have

(9) $$\boxed{[\alpha] \leq \alpha < [\alpha] + 1 \text{ and } \alpha - 1 < [\alpha] \leq \alpha}$$

For example

$$[\sqrt{10}] = 3, \quad \text{because } 3^2 < 10 < 4^2, 3 < \sqrt{10} < 4$$

$$[e] = 2, \quad e = 1 + 1 + \frac{1}{1!} + \frac{1}{2!} + \cdots, \quad 2 < e < 3$$

If $\pi$ is the area of the unit disc, then

$$3 < \pi < 4, \quad [\pi] = 3$$

Since $-4 < -\pi < -3$, we have $[-\pi] = -4$

In the division algorithm we saw that

$$a = q.m + r, m > 0, \quad 0 \leq r < m$$

Thus

$$\frac{a}{m} = q + \frac{r}{m}, \quad 0 \leq \frac{r}{m} < 1$$

This means that

(10) $$q = \left[\frac{a}{m}\right]$$

**Theorem 2.** *If $n$ is a positive integer, then* $\left[\frac{\alpha}{n}\right] = \left[\frac{[\alpha]}{n}\right]$

*Remark:* Since, by definition, $[\alpha]$ is the greatest integer not exceeding $\alpha$, we must have:

$$a \textit{ integer}, \quad a \leq \alpha \Rightarrow a \leq [\alpha]$$

**Proof of Theorem 2:** Let $k = \left[\frac{\alpha}{n}\right]$, then

$$k \leq \frac{\alpha}{n} < k+1, \quad kn \leq \alpha < (k+1)n$$

In view of the Remark, $kn \leq \alpha$ implies $kn \leq [\alpha]$; accordingly

$$kn \leq [\alpha] < (k+1)n$$

$$k \leq \frac{[\alpha]}{n} < k+1$$

which means that $k = \left[\frac{[\alpha]}{n}\right]$ ■

## § 2. DIVISIBILITY

Given two integers a and b, with $b > 0$, we say that a is a multiple of b, if there exists an integer q such that a = q.b; we also say that b *divides* a.

Notation: $b|a$

*Caution:* Division by 0 is ALWAYS excluded. If $b$ does not divide $a$, we write $b \dagger a$. Since $a = 1 \cdot a$ we always have $1|a$, and for $a \neq 0$, $a|a$. If

$$b|a, \quad \text{but } b \neq 1 \text{ and } b \neq a$$

we call $b$ a *proper* divisor of $a$.

**Theorem 3**

$1^0$ *$d_1|a$ and $d_2|d_1$ $(d_1 \neq 0, d_2 \neq 0)$ imply $d_2|a$*

$2^0$ *If $d|a$ then $kd|ka$* $(d \neq 0, k \neq 0)$

$3^0$ *If $d|x$ and $d|y$, then for any a, b* $d|ax + by$

This follows immediately from the definitions.

An integer $p > 1$ is called a *prime* number if it has no proper divisors, in other words:

$$p \text{ prime} \iff p > 1, \text{ and the only divisiors of p are the two divisors 1 and } p$$

For example, 2 is a prime; every prime $p \neq 2$ must be odd. The prime numbers $< 100$ are given by

2, 3, 5, 7, 11, 13, 17, 19, 23, 29, 37, 41, 43, 47, 53, 59, 61, 67, 71, 73, 79, 83, 89, 97.

**Theorem 4** Every integer $a > 1$ can be expressed as a product of primes.

**Proof:** For $a = 2$, the statement is valid since 2 is a prime, we prove the theorem by induction; suppose the assertion proved for $2, 3, \ldots, a-1$.

If a itself is a prime, there is nothing further to prove. Otherwise, by definition, a has proper divisors and so has a factorization.

(2) $$a = a_1 a_2, \quad 1 < a_1 < a \quad 1 < a_2 < a$$

By hypothesis of induction $a_1$ and $a_2$ can be expressed as products of primes; so is $a$ such a product because of (2). ∎

We call a number $> 1$, which is not a prime, a composite number.

**Theorem 5** *There exist infinitely many primes*

**Proof:** We show that there always exists a prime which is not contained in any given finite set of primes.

Let $A = \{p_1, p_2, \ldots, p_n\}$ be a set of primes. Let

$$a = 1 + p_1 p_2 \ldots p_n$$

If $a$ itself is a prime we have already a prime not in $A$. If $a$ is not a prime, by Theorem 4, $a$ has a prime factor $q$. We claim that $q \neq p_1, p_2, \ldots p_n$. This is obvious because $p_1, p_2 \ldots, p_n$ do not divide $a$. Hence $q$ is not in $A$. ∎

**Example**

(i) $2048 = 2.1024 = 2^2.512 = 2^3.256 = 2^3.16^2 = 2^3.2^8 = 2^{11}$

(ii) $756 = 2^2.3^3.7$

(iii) $10725 = 3.5^2.11.13$

Arranging the prime factors of $a$ in increasing order we can present the factorization of $a$ in prime factors as follows:

$$a = p_1^{\alpha_1} p_2^{\alpha_2} \ldots p_k^{\alpha_k}, \quad p_1 \ldots p_k \text{ distinct primes,}$$
$$\alpha_1 > 0, \ldots, \alpha_k > 0$$

We call this the *standard* or the *canonical* factorization of $a$.

*Remarks*

1) The most accurate table of prime numbers is by *D.H. LEHMER* List of prime numbers from 1 upto

10, 006, 721

Carnegie Institution, Washington (1914).

2) The following integer with 39 digits is a primes number.

$$2^{127} - 1 = 1701, 41183, 46046, 92317, 31687,$$
$$30371, 58841, 05727$$

3) The number

$$180(2^{127}-1)^2+1$$

has 79 digits, and is a prime number

4) The largest known prime so far is

$$2^{44497}-1$$

which has 13395 digits.

5) The following number

$$\begin{aligned}2^{257}-1 = {} & 231,58417,84746,32390,84714,\\ & 19700,17375,81570,65399,\\ & 69331,28112,80789,15168,\\ & 01582,62592,79871\end{aligned}$$

has 78 digits, and is known to be composite. But its prime factorization is not known. The proof of these facts needs special methods and computing machines.

6) *Loo Keng Hua:*

It is not difficult to construct a large prime number but it is not easy to factorize a large composite integer.

For example, it only takes 45 seconds computing time to find the first prime $> 2^{200}$: namely,

$$2^{200}+235,$$

a number with 61 digits.

But the computing time required to factorize a product of two primes, each with 61 digits, exceeds 4 million million years.

**Example 2.** Prove that, for $n > 1$, $H_n = 1+\frac{1}{2}+\ldots+\frac{1}{n}$ is never an integer.

**First Proof.** Fix $k$ such that $2^k \le n < 2^{k+1}$; thus, for $n > 1$, i.e. $n \ge 2$ we must have $2^{k+1} > n \ge 2$, so $2^k > 1$, $k \ge 1$.

Put

(3) $\quad P_n = 2^k \cdot Q_n,$ where $Q_n = \Pi b, \quad (b,2)=1, 1 \le b \le n$

We verify that $\frac{P_n}{\alpha}$ is an integer, $(a=1,\ldots,n)$. For this purpose, write $\alpha$ in the form $\alpha = 2^\alpha \cdot a', (a',2)=1$. Since

$$2^\alpha \le 2^\alpha a' = a \le n < 2^{k+1}, \text{ i.e.} 2^\alpha < 2^{k+1}$$

We must have $\alpha \leq k$. But then $2^\alpha | 2^k$. Since $a' | Q_n$ it follows that $a = 2^\alpha \cdot a' | P_n$.

Accordingly

(4) $A_n = P_n H_n = P_n(1 + \frac{1}{2} + \ldots + \frac{1}{n}) = \sum_{a=1}^{n} \frac{P_n}{a}$ is an *integer*

Let $G_n = \sum \frac{1}{b}$, and $Q_n = \Pi\, b \cdot 1 \leq b \leq n, (b, 2) = 1$. Then $Q_n G_n$ is an *integer*.

We have

$$H_n = \left(1 + \frac{1}{3} + \frac{1}{5} + \ldots\right) + \left(\frac{1}{2} + \frac{1}{4} + \frac{1}{6} + \ldots\right)$$

so that

(5) $$\boxed{H_n = G_n + \frac{1}{2} H_{[\frac{n}{2}]}}$$

Since $2^k \leq n < 2^{k+1}$ we have $2^{k-1} \leq \left[\frac{n}{2}\right] < 2^k$; accordingly, from (4)

$$A_n = 2^k Q_n H_n, \quad A_{[\frac{n}{2}]} = 2^{k-1} Q_{[\frac{n}{2}]} H_{[\frac{n}{2}]}$$

Also

$$Q_n = Q_{[\frac{n}{2}]} \cdot Q'_n \text{ where } Q'_n = \Pi b,\ (b, 2) = 1,\ \left[\frac{n}{2}\right] < b \leq n.$$

Writing (5) in terms of $A_n$, we obtain

(6) $$A_n = Q'_n A_{[\frac{n}{2}]} + 2^k Q_n G_n$$

Since, for $n > 1$, we have $k \geq 1$, and since $Q_n G_n$ is an integer we see from (6) that

(7) $$A_n \equiv Q'_n A_{[\frac{n}{2}]} \pmod{2}, \quad n \geq 2$$

Hence follows, at once, by induction on $n$:

(8) $$\boxed{A_n \equiv Q_n \pmod{2}} \quad (n \geq 2)$$

In fact, when $n = 2$, we have $k = 1$, $Q_2 = 1$, $A_2 = 2(1 + \frac{1}{2})$ so that (8) is valid for $n = 2$. Assuming (8) for values $< n$, we obtain from (7)

$$A_n \equiv Q'_n A_{[\frac{n}{2}]} \equiv Q'_n Q_{[\frac{n}{2}]}, \text{ by inductive hypothesis}$$

Since $Q_n = Q_{[\frac{n}{2}]} \cdot Q'_n$ we obtain the congruence (8).

Suppose $n > 1$, and $H_n$ is an integer. Then $A_n = 2^k Q_n H_n$ will be an *even* integer (since $k \geq 1$), i.e. $A_n \equiv 0 \pmod{2}$. But then (8) implies $Q_n \equiv 0 \pmod{2}$. But this is impossible. Consequently $H_n$, for $n > 1$, is never an integer. ■

**Second Proof.** We saw above that, if we fix $k$ such that $2^k \leq n < 2^{k+1}$, every $a$, $(1 \leq a \leq n)$, can be *uniquely* written in the form $a = 2^\alpha \cdot a'$, $(a', 2) = 1$, where $0 \leq \alpha \leq k$. Here equality $\alpha = k$ holds provided

$$2^k \cdot a' = a \leq n < 2^{k+1}, \text{ i.e. } a' < 2,$$

and since $a'$ is odd, we must have $a' = 1$. Accordingly

$$(9) \quad \left.\begin{array}{c} 1 \leq a \leq n \\ a \neq 2^k \end{array}\right\} \Rightarrow \begin{cases} a = 2^\alpha a', (2, a') = 1, \text{ with} \\ 0 \leq \alpha \leq k-1, \textit{thus } k - \alpha \geq 1 \end{cases}$$

$$\Rightarrow \frac{P_n}{a} = \frac{2^k Q_n}{2^\alpha a'} = 2^{k-\alpha} \cdot \left(\frac{Q_n}{a'}\right) \equiv 0 \pmod{2},$$

since $a' | Q_n$ and $k - \alpha \geq 1$

Consequently

$$P_n H_n = \sum_{a=1}^{n} \frac{P_n}{a} = \frac{P_n}{2^k} + \sum_{\substack{1 \leq a \leq n \\ a \neq 2^k}} \frac{P_n}{a} =$$

$$= Q_n + \{\text{sum of integers } \equiv 0 \pmod{2},\} \quad \text{[See (9)]}$$

which yields the congruence (8):

$$A_n = P_n H_n \equiv Q_n \pmod{2}$$

The proof can now be completed as before. ■

**Example 3** Prove that, for $n > 1$,

$$G_n = \sum \frac{1}{a}, \quad (a, 2) = 1, \quad 1 \leq a \leq n$$

is never an integer.

**First Proof.** The set of odd numbers can be divided into two different subsets: (i) $a$ odd, $a \equiv 0 \pmod{3}$, and (ii) $b$ odd, $b \not\equiv 0 \pmod{3}$, i.e. $b \not\equiv 0 \pmod{2}$, and $b \not\equiv 0 \pmod{3}$, or what is the same $(6, b) = 1$. We can write

$$G_n = \sum_{1 \leq 3a \leq n} \frac{1}{3a} + \sum_{\substack{1 \leq b \leq n \\ (6,b)=1}} \frac{1}{b}$$

which yields

$$(10) \qquad \boxed{G_n = \frac{1}{3} G_{[\frac{n}{3}]} + B_n} \quad \text{with } B_n = \sum_{\substack{1 \le b \le n \\ (6,b)=1}} \frac{1}{b}$$

Fix $k$ such that $3^k \le n < 3^{k+1}$. Since $n$ is odd and $> 1$, i.e. $n \ge 3$ we must have $k \ge 1$. Put

$$P_n = 3^k Q_n \quad \text{with} \quad Q_n = \Pi\, b, 1 \le b \le n, \ (b, 6) = 1$$

For each given $a, (a, 2) = 1, 1 \le a \le n$, we can *uniquely* write $a = 3^\alpha \cdot a'$, with $(6, a') = 1$. Since

$$3^\alpha \le 3^\alpha \cdot a' = a \le n < 3^{k+1}, \text{ thus } 3^\alpha < 3^{k+1}$$

We must have $\alpha \le k$. But then

$$a = 3^\alpha \cdot a' | 3^k \cdot Q_n = P_n \text{ since } \alpha \le k \text{ and } a' | Q_n$$

Accordingly

$$(11) \qquad A_n = P_n G_n = \sum_{\substack{1 \le a \le n \\ (2,a)=1}} \frac{P_n}{a} \text{ is an } \textit{integer}$$

Since $3^k \le n < 3^{k+1}$ we have $3^{k-1} \le [\frac{n}{3}] < 3^k$ so that

$$A_n = P_n G_n = 3^k \cdot Q_n G_n \text{ and } A_{[\frac{n}{3}]} = 3^{k-1} \cdot Q_{[\frac{n}{3}]} G_{[\frac{n}{3}]}$$

Also we have

$$Q_n = Q_{[\frac{n}{3}]} \cdot Q'_n \text{ where } Q'_n = \prod_{(b,6)=1} b, \quad [\frac{n}{3}] < b \le n$$

We can therefore obtain from (10):

$$A_n = 3^{k-1} Q_n G_{[\frac{n}{3}]} + 3^k Q_n B_n = Q'_n A_{[\frac{n}{3}]} \equiv 0 \pmod 3,$$

when $n \ge 3$ is odd and $k \ge 1$

since $Q_n B_n$ is an integer.

i.e.

$$(12) \qquad \boxed{A_n \equiv Q'_n A_{[\frac{n}{3}]} \pmod 3, \quad n \text{ odd}, \ \ge 3}$$

Using (12), by induction on $n$, we can at once derive the congruence

$$(13) \qquad \boxed{A_n \equiv Q_n \pmod 3}$$

For $n = 3$, one has $k = 1$, $Q_3 = 1$, $P_3 = 3$, $A_3 = 3^1 \cdot 1 \cdot (1 + \frac{1}{3}) = 4 \equiv 1 \pmod 3$ so that (13) holds for $n = 3$. Assume (13) for values

$< n$, thus, in particular:, $A_{[\frac{n}{3}]} \equiv Q_{[\frac{n}{3}]} \pmod 3$. Then, from (12),

$$A_n \equiv Q'_n A_{[\frac{n}{3}]} \equiv Q'_n Q_{[\frac{n}{3}]} = Q_n \pmod 3$$

which completes the induction and establishes (13),

Suppose $n > 1$ and $G_n$ is an integer, Then $k \geq 1$ and $A_n = 3^k Q_n G_n \equiv 0 \pmod 3$, so that (13) implies $Q_n \equiv 0 \pmod 3$ which, however, is impossible. Consequently, for $n > 1, G_n$ can never be an integer. ■

**Second Proof.** Fix $k$ such that $3^k \leq n < 3^{k+1}$; since $n \geq 3$ we must have $k \geq 1$. For each given odd $a$ we can uniquely put $a$ in the form $a = 3^\alpha a'$ with $(6, a') = 1$

We then have

$$3^\alpha \leq 3^\alpha a' = a \leq n < 3^{k+1}, \text{ thus } 3^\alpha < 3^{k+1}$$

whence follows $\alpha \leq k$. Herein equality $\alpha = k$ is possible only if $3^k \cdot a' \leq n < 3^{k+1}$, i.e. $a' < 3$ and since $(a', 6) = 1$, we must have $a' = 1$. Accordingly

(13) $1 \leq a \leq n, (2, a) = 1, a \neq 3^k \Rightarrow 0 \leq \alpha \leq k-1$, thus $k - \alpha \geq 1$

$$\Rightarrow a = 3^\alpha a', \quad (6, a') = 1,$$

$$\frac{P_n}{a} = \frac{3^k \cdot Q_n}{3^\alpha \cdot a'} = 3^{k-\alpha} \cdot \left(\frac{Q_n}{a'}\right)$$

$$\Rightarrow \frac{P_n}{a} \equiv 0 \pmod 3, \text{ since } k - \alpha \geq 1 \text{ and } a' | Q_n$$

Consequently

$$P_n G_n = \sum_{\substack{a=1 \\ (2,a)=1}}^{n} \frac{P_n}{a} = \frac{P_n}{3^k} + \sum_{\substack{1 \leq a \leq n \\ a \neq 3^k}} \frac{P_n}{a}$$

$$= Q_n + \sum (\text{integers } \equiv 0 (\text{mod } 3)),$$

[See (13)]

which yields the desired congruence (12):

$$P_n G_n \equiv Q_n \pmod 3$$

The proof is now completed as before. ■

**Example 4.** Show that, for any positive integer $n$

$$[x]+[x+\frac{1}{n}]+\ldots+[x+\frac{n-1}{n}]=[nx] \tag{10}$$

Let $f(x)=\sum_{k=0}^{n-1}\left[x+\frac{k}{n}\right]$. We can write

$$x=N+\theta \text{ with } N=[x] \text{ and } 0\leq\theta<1$$

Then $[x+\frac{k}{n}]=N+[\theta+\frac{k}{n}]$, so that

$$f(x)=nN+f(\theta)$$

Accordingly it is enough to prove that

$$f(\theta)=[n\theta]$$

Let $r=[n\theta]$; since $0\leq\theta<1$, we have $0\leq r<n$, thus

$$r\leq n\theta<r+1, \frac{r}{n}\leq\theta<\frac{r+1}{n},$$

$$\theta-\frac{r+1}{n}<0\leq\theta-\frac{r}{n}$$

$$\theta+\frac{n-r-1}{n}<1\leq\theta+\frac{n-r}{n}$$

In this manner

$$0\leq\theta<\theta+\frac{1}{n}<\ldots<\theta+\frac{n-r-1}{n}<1;$$

$$1\leq\theta+\frac{n-r}{n}<\ldots<0+\frac{n-1}{n}<1+1=2$$

Consequently

$$f(\theta)=\sum_{k=0}^{n-r-1}[\theta+\frac{k}{n}]+\sum_{k=n-r}^{n-1}\left[\theta+\frac{k}{n}\right]$$

$$=0+r=r=[n\theta]$$

This completes the proof.

**Remark: On Example 4**

We observe that f(x) satisfies the functional equation:

$$f(x+\frac{1}{n})=f(x)+1 \tag{11}$$

In fact,

$$f(x+\frac{1}{n})=[x+\frac{1}{n}]+[x+\frac{2}{n}]+\ldots+[x+\frac{n-1}{n}]+[x+\frac{n}{n}],$$

and since $[x+\frac{n}{n}]=[x+1]=[x]+1$, (11) is proved.

In general, for $r = 1, 2, \ldots$

$$f(x + \frac{r}{n}) = f(x + \frac{r-1}{n}) + 1 = \ldots = f(x) + r$$

and

$$f(x) = f((x - \frac{r}{n}) + \frac{r}{n}) = f(x - \frac{r}{n}) + r,$$

$$f(x - \frac{r}{n}) = f(x) - r$$

Hence

(12) $\boxed{f(x + \frac{r}{n}) = f(x) + r}$ $\quad (r = 0, \pm 1, \pm 2, \ldots)$

Also

(13) $\boxed{f(x) = 0}$ for $0 \le x < \frac{1}{n}$

In fact, we have then:

$$0 \le x < x + \frac{1}{n} < \ldots < x + \frac{n-1}{n} < \frac{1}{n} + \frac{n-1}{n} = 1$$

so that (13) follows.

Given $x$, we write:

$$nx = D + \delta, \; D = [nx] \text{ and } 0 \le \delta < 1$$

Then $x = \dfrac{\delta}{n} + \dfrac{D}{n}$ so that $f(x) = f(\dfrac{\delta}{n}) + D$, in view of the relation (12

Since now $0 \le \dfrac{\delta}{n} < \dfrac{1}{n}$ we must have $f(\dfrac{\delta}{n}) = 0$. Consequently

$$f(x) = 0 + D = [nx],$$

which yields the identity (10). ■

**Example 5.** If $f(x) = x - [x] - \dfrac{1}{2}$, and $S_n(x) = \displaystyle\sum_{k=1}^{n}\left(2^n x - [2^n x + \frac{1}{2}]\right)$ show that $|S_n(x)| \le 1$ for all $x$.

First we observe that

$$[x] + [x + \frac{1}{2}] = [2x]$$

In fact, if $m = [x], \; x = m + \theta, \; 0 \le \theta < 1$, we have

$$[x] = m, [x + \frac{1}{2}] = [m + \theta + \frac{1}{2}] = m + [\theta + \frac{1}{2}], [2x] = [2m + 2\theta]$$
$$= 2m + [2\theta]$$

so that

$$[x]+[x+\frac{1}{2}]-[2x]=[\theta+\frac{1}{2}]-[2\theta]=\begin{cases}0-0 \text{ for } 0\le\theta<1/2\\ 1-1 \text{ for } \frac{1}{2}\le\theta<1\end{cases}$$

Hence, always, $[x]+[x+\frac{1}{2}]-[2x]=0$

Returning to our example, we have

$$f(x)=x-[x]-\frac{1}{2}$$

$$f(x+\frac{1}{2})=x-[x+\frac{1}{2}]$$

so that $f(x)+f(x+\frac{1}{2})=2x-[x]-[x+\frac{1}{2}]-\frac{1}{2}$

$$=2x-[2x]-\frac{1}{2}$$

$$=f(2x)$$

Accordingly we obtain, for all $x$,

$$\boxed{f(x+\frac{1}{2})=f(2x)-f(x)}$$

Hence follows:

$$f(2^kx+\frac{1}{2})=f(2^{k+1}x)-f(2^kx)$$

Putting $k=1,2,\ldots,n$ here, and adding the equations so obtained we find

$$S_n(x)=\sum_{k=1}^{n}f(2^kx+\frac{1}{2})=f(2^{n+1}x)-f(2x)$$

But $[x]\le x<[x]+1$, $-\frac{1}{2}\le x-[x]-\frac{1}{2}<\frac{1}{2}$, thus $-\frac{1}{2}\le f(x)<\frac{1}{2}$., $|f(x)|\le\frac{1}{2}$. Consequently $|S_n(x)|\le 1$ for all $x$. ■

**Remark.** Similarly, we can verify that

$$\boxed{\sum_{k=1}^{n}\left[\frac{x}{2^k}+\frac{1}{2}\right]=[x]} \quad \text{for } 1\le x<2^n.$$

## § 3. GREATEST COMMON DIVISOR AND LEAST COMMON MULTIPLE

Given two positive integers $a$ and $b$ we have by the Division Algorithm (§1, Th.1, See. P.1)

$$
\begin{array}{lll}
 & a = bq_1 + r_2, & 0 < r_2 < b \\
 & b = r_2q_2 + r_3, & 0 < r_3 < r_2 \\
 & r_2 = r_3q_3 + r_4, & 0 < r_4 < r_3 \\
(1) & \vdots & \\
 & r_{n-2} = r_{n-1}q_{n-1} + r_n, & 0 < r_n < r_{n-1} \\
 & r_{n-1} = r_nq_n, & r_{n+1} = 0
\end{array}
$$

The sequence $\{r_k\}$ is a decreasing sequence of positive integers:

$$b > r_2 > r_3 \cdots > r_{n-1} > r_n$$

The process terminates with $r_{n+1} = 0$

If $d$ is a divisor of all the integers $a_1, a_2, \ldots a_n$, $d$ is said to be a common divisor. Since every number has only a finite number of divisors, the set of common divisors of $a_1, \ldots, a_n$ is a finite set. The maximum in this set is called the *greatest common divisor (g.c.d.)* of $a_1, \ldots a_n$. We write

$$g = (a_1, a_2, \ldots, a_n)$$

for the g.c.d. of $a_1, \ldots, a_n$. Two numbers $a$ and $b$ are said to be *coprime* or *relatively prime* if $(a, b) = 1$, thus

$$(a, b) = 1 \Leftrightarrow a \text{ and } b \text{ have no common divisors except } 1.$$

**Theorem 6.** *If* $b|a$, *then* $(a, b) = b$

**Proof** Let $g = (a, b)$, then $g$ being a common divisor of $a$ and $b$, $g|b$. Conversely, since $b|a$ and $b|b$, $b$ is a common divisor of $a$ and $b$. Consequently the set of common divisors of $a$ and $b$ coincides with the set of divisors of just $b$. Since the largest of the divisors of $b$ is $b$ itself, it follows that $g = b$ ■

**Theorem 7**

*If* $a_0 = q_0a_1 + a_2$, *then* $(a_0, a_1) = (a_1, a_2)$

**Proof** If $d$ is a common divisor of $a_0$ and $a_1$, i.e.

$$d|a_0, \quad d|a_1$$

then, [See Theorem 3, $3^0$, P.5],

$$d|a_0 - q_0 a_1, \text{ i.e } d|a_2$$

Thus $d$ is a common divisor of $a_1$ and $a_2$. Conversely, again by the same theorem

$$d|a_1 \text{ and } d|a_2 \Rightarrow d|q_0 a_1 + a_2, \text{ i.e. } d|a_0$$

i.e. $d$ is a common divisor of $a_0$ and $a_1$. Accordingly the set of common divisors of $a_0$ and $a_1$ coincides with the set of common divisors of $a_1$ and $a_2$. In particular, the largest among the divisors of $a_0$ and $a_1$ must coincide with the largest among the divisors of $a_1$ and $a_2$. ■

Applying Theorem 7 to the chain of equations (1), (See P.13), we see that

$$(a, b) = (b, r_2) = (r_2, r_3) = \cdots = (r_{n-1}, r_n) =$$
$$= (r_n q_n, r_n) = r_n$$

Consequently we obtain

**Theorem 8**

$1^0$ *The set of common divisors of a and b coincides with the set of divisors of the g.c.d. of a and b.*

$2^0$ *The last non-zero remainder $r_n$ in the Division Algorithm (1) yields the g.c.d. of a and b.*

We note the property of g.c.d. of $a$ and $b$ as follows:

(2) $$d|a \text{ and } d|b \Longleftrightarrow d|(a, b)$$

**Theorem 9**

$1^0 (am, bm) = (a, b).m$

$2^0$ *If $d|a$ and $d|b$, then* $(\frac{a}{d}, \frac{b}{d}) = \frac{a, b}{d}$

$3^0$ *If $g = (a, b)$, then* $(\frac{a}{g}, \frac{b}{g}) = 1$

$4^0$ *If $(a, b) = 1$, then $(am, b) = (m, b)$*

$5^0$ *If $d|ab$ and $(d, a) = 1$, then $d|b$*

**Proof** $1^0$ Multiply each equation in the chain (1) by $m$. The resulting chain yields the Euclidean Algorithm for determining $(am, bm)$. Hence the sequence of remainders is

$$am, \quad bm, \quad r_2 m, \quad \ldots, \quad r_n m$$

The last non-zero term, namely $r_n m$ must be the g.c.d. of $am$ and $bm$, thus

$$(am, bm) = r_n.m = (a, b)m$$

$2^0$ If $d|a$ and $d|b$, put $a = da'$, $b = db'$. Then, by $1^0$,

$$(a, b) = (da', db') = d(a', b'), \text{ so } \frac{(a, b)}{d} = (a', b') = \left(\frac{a}{d}, \frac{b}{d}\right)$$

$3^0$ Put $g = (am, b)$, $g' = (m, b)$. Then

$$g|am \text{ and } g|b \Rightarrow g|am \text{ and } g|bm \Rightarrow g|(am, bm)$$
$$\Rightarrow g|(a, b)m, \text{ i.e. } g|m \text{ since } (a, b) = 1$$
$$\Rightarrow g|m \text{ and } g|b \Rightarrow g|(m, b) = g'$$

Conversely

$$g'|m \text{ and } g'|b \Rightarrow g'|am \text{ and } g'|b \Rightarrow g'|(am, b) = g$$

In this mannner $g|g'$ and $g'|g$, hence $g = g'$ ■

*Remark* If $p$ is a *prime*, then for any integer a we must always have

$$(p, a) = \begin{vmatrix} 1 & \text{for } p \dagger a \\ p & \text{for } p|a \end{vmatrix}$$

Then, by Theorem 9, $5^0$ we have the fundamental property of primes

$$(3) \qquad \boxed{\left.\begin{array}{c} p|ab, p \dagger a \\ p \text{ prime} \end{array}\right\} \Rightarrow p|b}$$

which can also be stated as

$$(4) \qquad \boxed{\left.\begin{array}{c} p \dagger a_1, \ldots, p \dagger a_n \\ p \text{ prime} \end{array}\right\} \Rightarrow p \dagger a_1 a_2 \ldots a_n}$$

If $M$ is a multiple of both $a$ and $b$, it is said to be a common multiple of $a$ and $b$. We can write

$$M = \begin{vmatrix} am \\ bm' \end{vmatrix}$$

If $g = (a, b)$, we put $A = \dfrac{a}{g}$, $B = \dfrac{b}{g}$ so that $(A, B) = 1$ Then

$$am = bm', \quad gAm = gBm', \quad Am = Bm'$$

It follows from the last relation that $B|Am$. Since $(B, A) = 1$ we have, by (3), $B|m$; $m = Bt$. Then $M = am = aBt = \dfrac{ab}{g}.t$

Hence every common multiple of $a$ and $b$ is of the form $\frac{ab}{g}.t$ i.e., is a multiple of $\frac{ab}{g}$. For $t = 1$, we obtain the least among the common multiples. The *Least common multiple (l.c.m.)* of $a$ and $b$ is denoted by $[a, b]$. Thus

(5) $$[a,b] = \frac{ab}{g}, g = (a,b)$$
$$ab = (a,b).[a,b]$$

**Theorem 10**

*The g.c.d.* $(a, b)$ *has the following properties:*

$1^0$ *there exist integers* $x_0$ *and* $y_0$ *such that*

$$(a,b) = ax_0 + by_0$$

$2^0$ *For all integers* $x$ *and* $y$ *one always has*

$$(a,b)|ax + by$$

**Proof:** We have by the Euclidean Algorithm [§1, Th.1, See P.1]

(6) $$\begin{aligned} a &= bq_1 + r_2 \\ b &= r_2q_2 + r_3 \\ &\vdots \\ r_{n-2} &= r_{n-1}q_{n-1} + r_n \\ r_{(n-1)} &= r_nq_n \end{aligned} \qquad r_n = g \text{ where we put } g = (a,b)$$

Going backwards in the equations (6) we can solve for $r_n$:

$$\begin{aligned} r_n &= r_{n-2} - r_{n-1}q_{n-1} \\ r_{n-1} &= r_{n-3} - r_{n-2}q_{n-2} \\ &\vdots \end{aligned}$$

Ultimately $r_n$ is obtained as a linear combination of $a$ and $b$, i.e. $\ni$ integers $x_0$, $y_0$ with

$$(a,b) = g = r_n = ax_0 + by_0$$

This proves $1^0$. Next $g|a$, $g|b$ so $g|ax$, $g|by$ whence $g|ax + by$. This proves $2^0$.

*Corollary*

If $a$ and $b$ are coprime, then there exist integers $x_0$ and $y_0$ such that $ax_0 + by_0 = 1$.

This follows at once from the theorem since we now have $(a, b) = 1$

We have already seen (§2, Th. 4, See P.5) that every integer $a > 1$ can always be expressed as a product of primes. It is a deeper fact that this factorization is essentially unique. This uniqueness of factorisation stems from the basic property of prime numbers, namely: (see 4), P.16).

(7) $\boxed{p \text{ prime}, p|a_1, a_2, \ldots a_k \Rightarrow p| \text{ at least one of } a_1, \ldots, a_k}$

In particular, if $a_1, a_2, \ldots, a_k$ are themselves primes, then $p$ must coincide with one of them.

**Theorem 11**

*Every natural number $> 1$ can be resolved into a product of prime factors; further, except for order of factors, this factorization is unique.*

**Proof:** Suppose $n > 1$ and that $n$ has two canonical resolutions into prime factors:

$$n = \left| \begin{array}{ll} p_1^{\alpha_1} \ldots p_1^{\alpha_r}. & p_1 < \ldots < p_r \text{ primes} \\ \\ q_1^{\beta_1} \ldots q_s^{\beta_s}, & q_1 < \ldots < q_s \text{ primes} \end{array} \right.$$

As we just now remarked in (7) we see that

$$\text{each } p_i = a \text{ certain } q_j \text{ and conversely}$$

Therefore $r = s$. Since $p_1 < \ldots < p_r$ and $q_1 < \ldots < q_s$ we must have

$$p_1 = q_1, \ldots, p_r = q_r \qquad (r = s)$$

Also we must have the same exponents:

$$\alpha_1 = \beta_1, \ldots, \alpha_r = \beta_r \qquad (r = s)$$

In fact if, for instance $\alpha_1 > \beta_1$, then from

$$p_1^{\alpha_1} \ldots p_r^{\alpha_r} = p_1^{\beta_1} \ldots p_r^{\beta_r}$$

We can cancel the common factor $P_1^{\beta_1}$ and obtain

$$p_1^{\alpha_1 \beta_1} p_2^{\alpha_2} \ldots p_r^{\alpha_r} = p_2^{\beta_2} \ldots p_r^{\beta_r}$$

This is impossible since the left side is divisible by $p_1$ while the right side is *not* divisible by $p_1$ [See (7)].

## § 4. SIMPLE CONTINUED FRACTIONS

Let $\alpha$ be any real number. Denote by $a_0$ the integral part of $\alpha$:

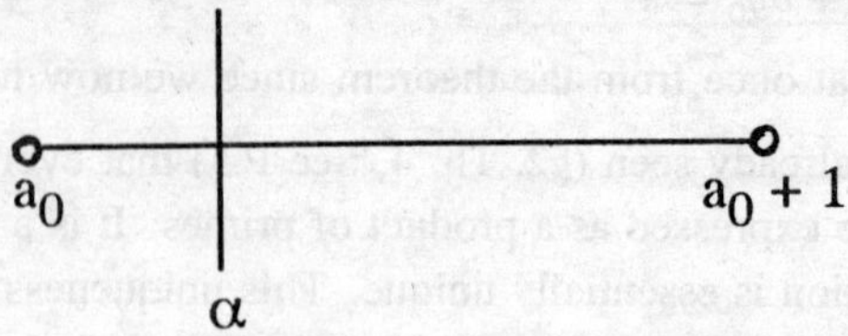


$$a_0 = [\alpha], \quad a_0 \le \alpha < a_0 + 1$$

Then $0 \le \alpha - a_0 < 1$. Suppose $\alpha$ is **not** an integer, then $0 < \alpha - a_0 < 1$, so that if we put

$$\alpha_1 = \frac{1}{\alpha - a_0}, \text{ then } \alpha = a_0 + \frac{1}{\alpha_1}, \quad (\alpha_1 > 1)$$

Proceeding with $\alpha_1$ as before with $\alpha$, we can write

$$a_1 = [\alpha_1], \quad \alpha_1 = a_1 + \frac{1}{\alpha_2}, \quad (\alpha_2 > 1)$$

We can continue in this manner as long as $\alpha_1, \alpha_2, \ldots, \alpha_k$, are not integers: the process terminates as soon as we arrive at $[\alpha_N]$ which is an integer, in which case $[\alpha_N] = a_N$ and we get

(1)
$$\alpha = a_0 + \cfrac{1}{a_1 + \cfrac{1}{a_2 + \ddots \cfrac{1}{a_{N-1} + \cfrac{1}{a_N}}}}$$

The expression in (4.1) is known as a *finite simple continued fraction.* We observe that the right side of (1) is a *rational* number.

For brevity, instead of (1), we agree to write

$$a_0 + \frac{1}{a_1+}\frac{1}{a_2+}\ldots\frac{1}{+a_N} \text{ or } [a_0, a_1, \ldots, a_N]$$

We wish to point out an intimate connection between the Division Algorithm and finite simple continued fractions. We illustrate this by the following example.

**Example 1.** $\alpha = \dfrac{100}{63}$

```
63 | 100  (1
      63
   ----
    37 | 63  (1
         37
       ----
        26 | 37  (1
             26
           ----
            11 | 26  (2
                 22
               ----
                 4 | 11  (2
                      8
                   ----
                     3 | 4  (1
                         3
                       ----
                         1 | 3  (3
                             3
                           ----
                             0
```

$$\frac{100}{63} = 1 + \frac{37}{63}$$
$$\frac{63}{37} = 1 + \frac{26}{37}$$
$$\frac{37}{26} = 1 + \frac{11}{26}$$
$$\frac{26}{11} = 2 + \frac{4}{11}$$
$$\frac{11}{4} = 2 + \frac{3}{4}$$
$$\frac{4}{3} = 1 + \frac{1}{3}$$
$$\frac{3}{1} = 3 + 0$$

(100, 63) =1

$$\frac{100}{63} = [1, 1, 1, 2, 2, 1, 3] \tag{2}$$

*Definition 1.* Given the continued fraction $\alpha = [a_0, a_1, \ldots, a_N]$ we define the sequence $C_0, C_1, \ldots C_N$ by

$$C_0 = a_0, C_1 = [a_0, a_1], C_2 = [a_0, a_1, a_2], \ldots, C_N = [a_0, a_1, \ldots, a_N]$$

In this manner

$$C_0 = a_0$$
$$C_1 = \frac{a_1 a_0 + 1}{a_1}$$
$$C_2 = \frac{a_2 a_1 a_0 + a_2 + a_0}{a_2 a_1 + 1}$$
$$\vdots$$

We call $C_n$ the $n$th convergent $(n = 0, 1, 2, \ldots, N)$

**Theorem 1**

*Define the sequences $\{p_n\}$ and $\{q_n\}$ by the recurrence relations*

$$\boxed{\begin{aligned} p_n &= a_n p_{n-1} + p_{n-2} \\ q_n &= a_n q_{n-1} + q_{n-2} \end{aligned}} \quad \text{with} \begin{cases} p_0 = a_0, & p_1 = a_1 a_0 + 1 \\ q_0 = 1, & q_1 = a_1 \end{cases} \tag{3}$$

*Claim:*

(4) $$C_n = [a_0, a_1, \ldots, a_n] = \frac{p_n}{q_n}$$

**Proof:** We prove the assertion (4) by induction on $n$. For $n = 0$, $C_0 = a_0 = \frac{a_0}{1} = \frac{p_0}{q_0}$; and for $n = 1$, $C_1 = a_0 + \frac{1}{a_1} = \frac{a_1 a_0 + 1}{a_1} = \frac{p_1}{q_1}$. We observe that $C_{n+1}$ results from $C_n$ by replacing in $C_n$ the last term $a_n$ by $a_n + \frac{1}{a_{n+1}}$, while leaving $a_0, a_1, \ldots, a_{n-1}$ unaltered. Hence, assuming (4) for $n$, we have

(5) $$C_n = \frac{p_n}{q_n} = \frac{a_n p_{n-1} + p_{n-2}}{a_n q_{n-1} + q_{n-2}}$$

Thus if in (5) on the right side we replace $a_n$ by $a_n + \frac{1}{a_{n+1}}$ –(this will not affect $p_{n-1}$, $p_{n-2}$, $q_{n-1}$, $q_{n-2}$, as they do not involve $a_n$ at all)–we obtain $C_{n+1}$. Hence

$$C_{n+1} = \frac{\left(a_n + \frac{1}{a_{n+1}}\right) p_{n-1} + p_{n-2}}{\left(a_n + \frac{1}{a_{n+1}}\right) q_{n-1} + q_{n-2}}$$

$$= \frac{(a_n p_{n-1} + p_{n-2}) + \frac{p_{n-1}}{a_{n+1}}}{(a_n q_{n-1} + q_{n-2}) + \frac{q_{n-1}}{a_{n+1}}}$$

$$= \frac{p_n + \frac{pn-1}{a_{n+1}}}{q_n + \frac{qn-1}{a_{n+1}}}, \text{ by (3)}$$

$$= \frac{a_{n+1} p_n + p_{n-1}}{a_{n+1} q_n + q_{n-1}}$$

$$= \frac{p_{n+1}}{q_{n+1}}, \text{ again by (3)}$$

This completes the induction and establishes the theorem.

**Theorem 2** *The convergents* $\frac{p_n}{q_n}$ *satisfy the relation*

(6) $$\boxed{p_n q_{n-1} - p_{n-1} q_n = (-1)^{n-1}} \qquad (n = 1, 2, \ldots)$$

**Proof:** Since we have

$$p_n q_{n-1} - p_{n-1} q_n = (a_n p_{n-1} + p_{n-2}).q_{n-1} - p_{n-1}(a_n q_{n-1} + q_{n-2})$$
$$= -(p_{n-1} q_{n-2} - p_{n-2} q_{n-1})$$

we can write

$$t_n = -t_{n-1} \text{ where } t_n = p_n q_{n-1} - p_{n-1} q_n$$

Hence follows

$$t_n = -t_{n-1} = +t_{n-2} = -t_{n-3} = \ldots = (-1)^{n-1} t_1$$

But $t_1 = p_1 q_0 - p_0 q_1 = (a_1 a_0 + 1).1 - a_0 a_1 = 1$ ■

We shall display the relation (6) also as:

(7) $$\begin{vmatrix} p_n & p_{n-1} \\ q_n & q_{n-1} \end{vmatrix} = (-1)^{n-1} \qquad (n = 1, 2, \ldots)$$

**Example 2** [BHASKARA (II), Lilâvati]

Solve in positive integers

$$\frac{100x + 90}{63} = y$$

We have found in Example 1, the simple continued fraction expansion for the rational number $\alpha = \frac{100}{63}$. Using the recurrence relations (3) we determine the sequence of convergents $\frac{p_n}{q_n}$. It is convenient to arrange the computations in a tabular form as follows:

| | $a_0$ | $a_1$ | $a_2$ | $a_3$ | $a_4$ | $a_5$ | $a_6$ |
|---|---|---|---|---|---|---|---|
| | 1 | 1 | 1 | 2 | 2 | 1 | 3 |
| $p_n$ | 1 | 2 | 3 | 8 | 19 | 27 | 100 |
| $q_n$ | 1 | 1 | 2 | 5 | 12 | 17 | 63 |

$$p_{N-1} = 27 \quad p_N = 100 \quad \text{(N = 6)}$$
$$q_{N-1} = 17 \quad q_N = 63$$

(8) $$\begin{vmatrix} p_N & p_{N-1} \\ q_N & q_{N-1} \end{vmatrix} = \begin{vmatrix} 100 & 27 \\ 63 & 17 \end{vmatrix} = -1 = (-1)^{N-1}$$

If we take

$$(x_0, y_0) = (q_{N-1}, p_{N-1}) = (17, 27)$$

the relation (8) becomes

$$100x_0 - 63y_0 = -1$$

Multiplying both sides by 90 we get

$$100x - 63y = -90$$

where

$$(x, y) = (90x_0, 90y_0) = (1530, 2430)$$

Put

$$(x_1, y_1) = (1530, 2430)$$

This yields one solution of the equation. In order to find all solutions $(x, y)$ we have

$$\frac{100x + 90}{63} = y, \quad \frac{100x_1 + 90}{63} = y_1$$

whence follows

$$100x - 100x_1 = 63y - 63y_1$$
$$\frac{x - x_1}{63} = \frac{y - y_1}{100} = t$$
$$(x, y) = (x_1 + 63t, y_1 + 100t)$$
$$= (1530 + 63t, 2430 + 100t)$$

As $t$ runs through all integers we obtain all the integer solutions of the given equation. In order to obtain solutions in positive integers we need only choose the integer $t$ such that $1530 + 63t$ and $2430 + 100t$ are positive.

**Theorem 3**

*Given: integers* $a > 0,\ \ b > 0$ *with* $(a, b) = 1$

*Claim* $1^\circ$ Every integer $> ab - a - b$ can be represented as

(9) $\qquad ax + by$ with $x \geq 0, \ \ y \geq 0$

$2^\circ ab - a - b$ cannot be so represented

**Proof:** $1^0$ Let $n$ be an integer with $n > ab - a - b$. We wish to solve for $x, y$ in the equation

$$n = ax + by$$

for non-negative integers $x, \ \ y$. Start with any particular solution $(x_0, \ \ y_0)$:

(10) $$n = ax_0 + by_0$$

Then

$$a(x - x_0) + b(y - y_0) = 0 \text{ or } \frac{x - x_0}{b} = \frac{y - y_0}{-a} = t$$

If that particular solution $(x_0, y_0)$ is itself such that $x_0 \geq 0,\ y_0 \geq 0$ then there is nothing further to prove. Otherwise we choose $t$ such that

$$(x, y) = (x_0 + bt, y_0 - at), \qquad x \geq 0, y \geq 0 \tag{11}$$

We show that such a choice of $t$ is possible. First choose $t$ so that

$$0 \leq y_0 - at < a, \quad i.e.\ 0 \leq y_0 - at \leq a - 1$$

Now we have

$$n = a(x_0 + bt) + b(y_0 - at) \leq a(x_0 + bt) + b(a - 1)$$

so that

$$ab - a - b < n \leq a(x_0 + bt) + b(a - 1)$$
$$(ab - a - b) - b(a - 1) < a(x_0 + bt)$$
$$-a < a(x_0 + bt)$$

Since $a > 0$ this implies

(12) $\quad x_0 + bt > -1$, thus, being an integer $x_0 + bt \geq 0$

We see from (11) and (12), that the choice of t has been made as required

$2^0$ *claim:* $ab - a - b$ is the *largest* integer which cannot be represented in the form

$$ax + by$$

with non-negative coefficients $x \geq 0,\ \ y \geq 0$.

Suppose, if possible, $ab - a - b = ax_1 + by_1$ with $x_1 \geq 0,\ y_1 \geq 0$ Then

$$ab = a(x_1 + 1) + b(y_1 + 1)$$

It follows that

$$a|b(y_1 + 1) \text{ and } b|a(x_1 + 1)$$

Since $(a, b) = 1$ this implies $a|y_1 + 1$ and $b|x_1 + 1$. Therefore

$$(y_1 + 1) \geq a \text{ and } (x_1 + 1) \geq b$$

Consequently

$$ab = a(x_1 + 1) + b(y_1 + 1) \geq ab + ba = 2ab$$

which is absurd. Our supposition is wrong. The theorem is proved. ■

In connection with Theorem 3 one is naturally led to seek a generalization to the case of more than two coprime numbers.

*PROBLEM:* Given three positive integers which are mutually coprime:

$a > 0,\ \ b > 0,\ \ c > 0$ with $(a, b) = (a, c) = (b, c) = 1$

Determine the largest integer which cannot be represented in the form $ax + by + cz$ with $x \geq 0$, $y \geq 0$, $z \geq 0$.

This is an unsolved problem.

## § 5. THE EULER $\varphi$-FUNCTION

*Definition 1.* We put $\varphi(n) = 1$, for $n = 1$, and for $n > 1$, $\varphi(n) =$ number of integers among $1, 2, \ldots, n$ that are coprime with the last integer $n$

| $n =$ | 1 | 2 | 3 | 4 | 5 | 6 | 7 | 8 | 9 | 10 |
|---|---|---|---|---|---|---|---|---|---|---|
| $\varphi(n) =$ | 1 | 1 | 2 | 2 | 4 | 2 | 6 | 4 | 6 | 4 |

We observe that, if $p$ is a prime, all the $p-1$ numbers $1, 2, \ldots, p-1$ are coprime with $p$, i.e.

(1) $\qquad \varphi(p) = p - 1$ whenever $p$ is a prime

We wish to compute $\varphi(n)$ for any integer $n > 1$. For this purpose we establish a general proposition which is frequently applied in number theory.

**Theorem 1. (COUNTING PRINCIPLE)**

*Given:* $N$ objects and properties $\alpha_1, \alpha_2, \ldots, \alpha_p$

$N_{\alpha_i}$ = number of objects with the property $\alpha_i$

$N_{\alpha_i\alpha_j}$ = number of objects with the properties $\alpha_i$ and $\alpha_j$

$N_{\alpha_i\alpha_j\alpha_k}$ = number of objects having all the three properties $\alpha_i$, $\alpha_j$ and $\alpha_k$

*Claim:* If

$\Phi(N) =$ the number of objects having **none** of the properties $\alpha_i, \ldots, \alpha_p$

then

$$\Phi(N) = N - \sum_i N_{\alpha_i} + \sum_{i,j} N_{\alpha_i\alpha_j} - \sum_{i,j,k} N_{\alpha_i\alpha_j\alpha_k} + \cdots + (-1)^p N_{\alpha_1\cdots\alpha_p}$$

**Proof:** Denote by $\alpha'$ the negation of the property $\alpha$, and by $S_\alpha$ the set of objects with the property $\alpha$. Thus $N_\alpha$ is the number of elements in $S_\alpha$. Hence

$$N = N_\alpha + N_{\alpha'}$$

Let $S$ denote the given set of objects: $S = S_\alpha + S_{\alpha'}$ (disjoint union) Put $S_{\alpha\beta} = S_\alpha \cap S_\beta$, i.e., $S_{\alpha\beta}$ in the set of objects having both the properties of $\alpha$ and $\beta$. Similarly, for example, $S_{\alpha'\beta}$ is the set of objects having the property $\beta$ but not $\alpha$. The number of elements in

$$S_\alpha,\ S_{\alpha'},\ S_{\alpha\beta},\ S_{\alpha'\beta},\ \ldots$$

are, respectively

$$N_\alpha,\ N_{\alpha'},\ N_{\alpha\beta},\ N_{\alpha'\beta},\ \ldots$$

We have the *disjoint* unions of sets:

$$S = \begin{cases} S_\alpha + S'_\alpha \\ S_\beta + S_{\beta'} \end{cases} S_{\alpha'\beta} + S_{\alpha'\beta'},\quad S_\beta = S_{\alpha\beta} + S_{\alpha'\beta}$$

Counting the number of elements we obtain, accordingly

(3) $$\begin{cases} N = N_\alpha + N'_\alpha \\ N'_\alpha = N - N_\alpha \end{cases} \begin{cases} N'_\alpha = N_{\alpha'\beta} + N_{\alpha'\beta'} \\ N_\beta = N_{\alpha\beta} + N_{\alpha'\beta} \end{cases}$$

Hence follows

$$N_{\alpha'\beta'} = N_{\alpha'} - N_{\alpha'\beta} = (N - N_\alpha) - (N_\beta - N_{\alpha\beta})$$

i.e.

(4) $$N_{\alpha'\beta'} = N - (N_\alpha + N_\beta) + N_{\alpha\beta}$$

It is clear from (3) and (4) that the formula (2) holds when there is only one property $\alpha$ or two properties $\alpha$ and $\beta$. The general argument is best illustrated by:

*Case* $p = 3$ Suppose we have *three* properties $\alpha$, $\beta$, $\gamma$. Partition the set $S_{\alpha'\beta'}$ into two subsets: (i) those with the property $\gamma$, and (ii) those without $\gamma$.

$$S_{\alpha'\beta'} = S_{\alpha'\beta'\gamma} + S_{\alpha'\beta'\gamma'} \text{ which implies}$$

$$N_{\alpha'\beta'} = N_{\alpha'\beta'\gamma} + N_{\alpha'\beta'\gamma'}$$

i.e.

(5) $$N_{\alpha'\beta'\gamma'} = N_{\alpha'\beta'} - N_{\alpha'\beta'\gamma}$$

Herein $N_{\alpha'\beta'} = N_{\alpha\beta}$ is computed in (4), and $N_{\alpha'\beta'\gamma}$ is the number of objects in $S_\gamma$ not having two properties $\alpha$ and $\beta$. Applying the case $p = 2$ to the set $S_\gamma$ instead of $S$, we obtain from (4):

$$N_{\alpha'\beta'\gamma} = N_\gamma - (N_{\alpha\gamma} + N_{\beta\gamma}) + N_{\alpha\beta\gamma}$$

From (5) and (6), using (4), we get

$$\begin{aligned} N_{\alpha'\beta'\gamma'} &= [N - (N_\alpha + N_\beta) + N_{\alpha\beta}] - [N_\gamma - (N_{\alpha\gamma} + N_{\beta\gamma}) + N_{\alpha\beta\gamma}] \\ &= N - (N_\alpha + N_\beta + N_\gamma) + (N_{\alpha\beta} + N_{\alpha\gamma} + N_{\beta\gamma}) - \\ &\quad - N_{\alpha\beta\gamma} \end{aligned}$$

which proves the theorem, when $p = 3$.

The general proof, by induction on the number of properties is now clear. ■

**Application to EULER $\varphi$ function:**

Given the integer $n > 1$, denote by $p_1, \ldots, p_k$ *all the distinct* prime factors of $n$

$$n = p_1^{\gamma_1} p_2^{\gamma_2} \ldots p_k^{\gamma_k}, \quad p_i \neq p_j \text{ primes}$$

Consider the set

$$S = \{1, \; 2, \; \ldots, \; n\}$$

and the properties $\alpha_1, \;\; \alpha_2, \;\; \ldots, \;\; \alpha_k$:

*Property* $\alpha_j$: $x \in S$ is divisible by the prime $p_j$ $(j = 1, \ldots, k)$

Since $(x, n) = 1 \Longleftrightarrow n$ is *not* divisible by any one of the $p_1, \cdots, p_k$

Hence

$$N_{\alpha_1' \ldots \alpha_k'} = \varphi(n)$$

$$N_{\alpha_1} = \frac{n}{p_1} = \text{the number of elements in } S \text{ divisible by } p_1 : \{p_1, 2p_1, \ldots, \frac{n}{p_1}.p_1\}$$

And so on. By the counting principle

$$\varphi(n) = n - \sum \frac{n}{p_i} + \sum \frac{n}{p_i p_j} - + \cdots + (-1)^k \frac{n}{p_1 p_2 \cdots p_k}$$

$$= n\left(1 - \frac{1}{p_1}\right)\left(1 - \frac{1}{p_2}\right) \cdots \left(1 - \frac{1}{p_k}\right)$$

We have accordingly obtained

**Theorem 2.** *If* $n = p_1^{\alpha_1} \cdots p_k^{\alpha_k}$, *then*

$$\varphi(n) = n\left(1 - \frac{1}{p_1}\right)\left(1 - \frac{1}{p_2}\right) \cdots \left(1 - \frac{1}{p_k}\right) \tag{7}$$

We shall also write this in the form

$$\frac{\varphi(n)}{n} = \Pi_{p|n}\left(1 - \frac{1}{p}\right) \quad (p \text{ prime}) \tag{8}$$

We shall now derive the basic multiplicative property of the $\varphi$ function.

**Theorem 3.** *If* $(m,n)=1$, *then*

$$\text{(9)} \qquad \boxed{\varphi(mn)=\varphi(m)\varphi(n)}$$

**Proof:** Let $m=p_1^{a_1}\cdots p_r^{a_r}$, $n=q_1^{b_1}\cdots q_s^{b_s}$ be the canonical factorizations. Herein

$$(m,n)=1 \Rightarrow p_i \neq q_j \quad (i=1,\cdots r;\ \ j=1,\cdots,s)$$

Hence the canonical factorization of $mn$ is

$$mn=p_1^{a_1}\cdots p_1^{a_r}q_1^{b_1}\cdots q_s^{b_s}$$

By (7) we have

$$\varphi(m)=m\left(1-\frac{1}{p_1}\right)\cdots\left(1-\frac{1}{p_r}\right),\ \ \varphi(n)=n\left(1-\frac{1}{q_1}\right)\cdots\left(1-\frac{1}{q_s}\right)$$

$$\varphi(mn)=mn\left(1-\frac{1}{p_1}\right)\cdots\left(1-\frac{1}{p_r}\right)\left(1-\frac{1}{q_1}\right)\cdots\left(1-\frac{1}{q_s}\right)$$

which at once yields (9).

**Theorem 4.**

If $\sigma(n)$ denotes the sum of all the divisors of $n$ then

$$\text{(10)} \qquad \boxed{\sigma(n)=\frac{p_1^{\alpha_1+1}-1}{p_1-1}\cdots\frac{p_k^{\alpha_k+1}-1}{p_k-1}}$$

$$(n=p_1^{\alpha_1}\cdots p_k^{\alpha_k})$$

**Proof:** If $d|n$, it must be of the form

$$d=p_1^{t_1}\cdots p_k^{t_k},\quad 0\le t_1\le\alpha_1,\cdots,0\le t_k\le\alpha_k$$

Hence

$$\sigma(n)=\sum_{d|n}d=\sum_{t_1,\cdots,t_k}p_1^{t_1},\cdots,p_k^{t_k}=$$

$$=(1+p_1+\cdots+p_1^{\alpha_1})(1+p_2+\cdots+p_2^{\alpha_2})\cdots(1+p_k+\cdots+p_k^{\alpha_k})$$

$$=\frac{p_1^{\alpha_1+1}-1}{p_1-1}\cdots\frac{p_k^{\alpha_k+1}-1}{p_k-1}$$

which proves (10)

*Corollary* If $(m,n)=1$, then $\sigma(mn)=\sigma(m)\sigma(n)$.

This follows at once from (10)

**Examples**

(i) $n=6,\quad \sigma(n)=1+2+3+6=12$

(ii) $n=28,\quad \sigma(n)=1+2+4+7+14+28=56$

A positive integer $n$ is called a *perfect* number if $\sigma(n) = 2n$
Thus 12 and 56 are perfect numbers.

The question whether there exists an *odd* perfect number is a famous unsolved problem.

## §6. THE MÖBIUS $\mu$—FUNCTION

*Definition 1.* The Möbius $\mu$ function, which is intimately related to the Euler $\varphi$ function, is defined as follows:

(i) $\mu(n) = 0$ whenever $n$ is divisible by the square of some prime (i.e. whenever $n$ has a squared factor)

(ii) If $n$ is square-free, i.e., $n = p_1 p_2 \cdots p_k$ is a product of *distinct* primes,

$$\mu(n) = \mu(p_1 \cdots p_k) = (-1)^k$$

(iii) $\mu(1) = 1$

For example:

$$\mu(p) = -1 \text{ for any prime } p$$
$$\mu(p^2) = 0 \text{ for any prime } p$$
$$\mu(8) = \mu(2^3) = 0$$
$$\mu(10) = \mu(2.5) = (-1)^2 = 1$$

If $n = p_1^{e_1} \cdots p_k^{e_k}$ is the canonical factorization

$$\mu(n) = 0 \text{ as soon as one of the } e_1, \cdots, e_k \text{ is } \geq 2$$

If $m = q_1^{l_1} \cdots q_j^{l_j}$ and $(m, n) = 1$, then it is clear that

(1) $$\boxed{\mu(mn) = \mu(m)\mu(n)}$$

i.e. $\mu$ is multiplicative

**Theorem 1**

$1^0$ *The product of two multiplicative functions is again multiplicative*

$2^0$ *If $m = p_1^{e_1} \ldots p_k^{e_k}$ and $f$ is any multiplicative function then*

(2) $$\sum_{d|n} f(d) = g(p_1)g(p_2) \cdots g(p_k)$$

where

(3) $$g(p_j) = 1 + f(p_j) + f(p_j^2) + \ldots + f(p_j^{e_j})$$

$$(j = 1, \ldots, k)$$

**Proof:** $1^0$ If $\alpha$ and $\beta$ are two multiplicative functions, and $\gamma = \alpha \cdot \beta$, then for $(m, n) = 1$

$$\alpha(mn) = \alpha(m)\alpha(n), \quad \beta(mn) = \beta(m)\beta(n)$$

so that

$$\begin{aligned}\gamma(mn) &= \alpha(mn)\beta(mn) = \alpha(m)\alpha(n).\beta(m)\beta(n) \\ &= \alpha(m)\beta(m).\alpha(n)\gamma(n) = \gamma(m)\gamma(n)\end{aligned}$$

Also $\gamma(1) = \alpha(1)\beta(1) = 1$.

$2^0$ From (2) and (3) we see that the product on the right side of (2) is a sum of terms of the form

$$f(p_1^{t_1})f(p_2^{t_2})\dots f(p_k^{t_k})$$
$$0 \le t_1 \le e_1, \dots, 0 \le t_k \le e_k$$

where each one of these occurs without omission. Since f is multiplicative, we get

$$g(p_1)\dots g(p_k) = \sum_{t_1=0}^{e_1} \dots \sum_{t_k=0}^{e_k} f(p_1^{t_1} p_2^{t_2} \dots p_k^{e_k})$$

But $p_1^{t_1} \dots p_k^{t_k}$ is a divisor of $n$, and all the divisors are obtained in this way; accordingly

$$g(p_1)\dots g(p_k) = \sum_{d|n} f(d) \blacksquare$$

**Theorem 2.** *If $n = p_1^{e_1} \dots p_k^{e_k}$ is the canonical factorization, and $f$ is any multiplicative function*

$$\boxed{\sum_{d|n} \mu(d) f(d) = (1 - f(p_1)) \dots (1 - f(p_k))} \quad (n > 1) \tag{4}$$

**Proof:** Since $\mu(d) = 0$ whenever $d$ has a squared factor, we need only retain in the sum only those terms with square free $d$; also (by definition of a multiplicative function) $f(1) = 1$, so that

$$\sum_{d|n} \mu(d)f(d) = \mu(1)f(1) + \sum_i \mu(p_i)f(p_i) + \sum \mu(p_ip_j)f(p_ip_j) + \cdots$$
$$= f(1) - \sum_i f(p_i) + \sum_{i,j} f(p_i)f(p_j) = \dots$$
$$+(-1)^k f(p_1)f(p-2) + \cdots f(p_k)$$
$$= (1 - f(p_1))(1 - f(p_2) \dots (1 - f(p_k) \blacksquare$$

If we take $f(n) \equiv 1$ for all $n$, we get from (4) the important property of the Möbius function:

$$\text{(5)} \qquad \boxed{\sum_{d|n} \mu(d) = \begin{vmatrix} 0 \text{ for } n > 1 \\ 1 \text{ for } n = 1 \end{vmatrix}}$$

We shall now derive a generalization of the Counting Principle (Th.1 § 5, See P. 25) which is convenient for applications. In the notation of that theorem, instead of merely counting the number of elements in the various sets, $S$, $S_\alpha$, $S_{\alpha\beta}\cdots$ we shall consider a real or complex-valued function defined on $S$, and put

$$F_{\alpha_i} = \sum_{x \in S_{\alpha_i}} f(x), \quad F_{\alpha_i\alpha_j} = \sum_{x \in S_{\alpha_i\alpha_j}} f(x), \ldots$$

and

$$F' = \sum_{x \in S'} f(x)$$

(In particular, when we take $f(x) \equiv 1$ for all $x$ in $S$, we have

$$F' = \Psi(N), \quad F_{\alpha_i} = N_{\alpha_i}, \quad F_{\alpha_i\alpha_j} = N_{\alpha_i\alpha_j}, \cdots$$

In fact, if $f$ is defined on any finite set $A = \{a_1, \ldots a_m\}$

$$\sum_{x \in A} f(x) = f(a_1) + \cdots + f(a_m)$$

and when $f(x) \equiv 1$ on $A$,

$$\begin{aligned} \sum_{x \in A} f(x) &= 1 + \cdots + 1 (m \text{ times }) \\ &= \text{the number of elements in } A \end{aligned}$$

The proof of the Counting Principle Theorem goes through for the present generalization with hardly any change. Accordingly we can state:

**Theorem 3. (Generalized Counting Principle)**

*Given: S is a finite set of objects, and properties of $\alpha, \beta, \gamma, \ldots$ of these objects.*

*$S_\alpha$ = the set of objects with the property $\alpha$*

*$S_{\alpha\beta}$ = the set of objects with both the properties $\alpha$ and $\beta$*

*$S_{\alpha\beta\gamma}$ = the set of objects with all the three properties $\alpha$, $\beta$, $\gamma$.*

*And so on. Further*

*$S'$ = the set of objects with none of these properties*

$$F_\alpha = \sum_{x \in S_\alpha} f(x), \quad F_{\alpha\beta} = \sum_{x \in S_{\alpha\beta}} f(x), \ldots \text{ and } F' = \sum_{x \in S'} f(x)$$

Then:

$$(6) \qquad \boxed{F' = \sum_{x \in S} f(x) - \sum_{\alpha} F_\alpha + \sum_{\alpha,\beta} F_{\alpha\beta} - \sum_{\alpha,\beta,\gamma} F_{\alpha\beta\gamma} + - \cdots}$$

*An application:*

Let $n = p_1^{e_1} \dots p_k^{e_k} (n > 1)$ be the canonical factorization of $n$.

Take $S = \{1, 2, \dots, n\}$, and let

$$S_i = \{a \in S, \quad a \text{ is divisible by } p_i\},$$

$$S_{ij} = \{a \in S, \quad a \text{ is divisible by } p_i \text{ and } p_j\}, \dots$$

$$\text{and } S' = \{a \in S, \quad a \text{ is not divisible by } p_1, p_2, \dots, p_k\}$$

$$\text{i.e. } a \in S' \iff (a, n) = 1$$

Thus

$$a \text{ has property } \alpha_i \text{ if and only if } p_i | a (i = 1, \dots, k)$$

Hence we have:

$$F' = \sum_{\substack{1 \le x \le n \\ (x,n)=1}}, f(x)$$

$$(7) \qquad F_{\alpha_i} = \sum_{p_i | x} f(x) = f(p_i) + f(2p_i) + \cdots + f\left(\frac{n}{p_i} \cdot p_i\right)$$

$$(i = 1, \dots k)$$

We are going to see that, in terms of the Möbius fucntion, the formula (6) assumes the following form:

$$(8) \qquad \boxed{\sum_{\substack{1 \le x \le n \\ (x,n)=1}} f(x) = \sum_{d|n} \mu(d) \left\{ f(d) + f(2d) + \cdots + f\left(\frac{n}{d} \cdot d\right) \right\}}$$

In fact

$$\sum_{x \in S} f(x) = f(1) + \cdots + f(n),$$

$$(9) \quad \sum_{x \in S_i} f(x) = f(p_i) + f(2p_i) + \cdots + f\left(\frac{n}{p_i} \cdot p_i\right),$$

$$\sum_{x \in S_{ij}} f(x) = f(p_i p_j) + f(2p_i p_j) + \cdots + f\left(\frac{n}{p_i p_j} \cdot p_i p_j\right), \cdots,$$

Since $\mu(d) = 0$ whenever $d$ is not square free we need only retain the terms corresponding to

$$d = 1, \quad p_1, \ldots, p_n, \quad p_1, p_2, p_3, \ldots, p_1, p_2, p_3, \ldots, \quad p_1, p_2, \cdots p_n$$

We see from (9) that (8) results at once from (6)

We state this separately as follows:

**Theorem 4.**

*Let*

$$(10) \qquad F_d(n) = \sum_{k=1}^{n/d} f(kd) \text{ and } F'(n) = \sum_{\substack{1 \le x \le n \\ (x,n)=1}} f(x)$$

*Then*

$$(11) \qquad \boxed{F'(n) = \sum_{d|n} \mu(d) F_d(n)}$$

**Example 1.** Prove that the sum of all the numbers not exceeding $n$ and relatively prime to $n$ is $\frac{n}{2}\varphi(n)$

Let $r_1, \ldots, r_h (h = \varphi(n))$ be the numbers in $\{1, 2, \ldots, n\}$ relatively prime to $n$. Then $n - r_1, \quad n - r_2, \quad \ldots, \quad n - r_h$ will be again the same numbers in a different order

$$\{r_1, \ldots, r_h\} = \{n - r_1, \ldots, n - r_h\}$$

$$\text{Hence } (r_1 + \cdots, +r_h) = (n - r_1) + \cdots + (n - r_h)$$

$$2(r_1 + \cdots + r_h) = n \cdot h = n \cdot \varphi(n)$$

$$(r_1 + \cdots + r_h) = \tfrac{n}{2}\varphi(n)$$

**Example 2.** Let $n > 1$

If $\sigma_m(n) = \sum_{\substack{1 \le x \le n \\ (x,n)=1}} x^m$ and $S_m(n) = 1^m + 2^m + \cdots + n^m$

Show that

$$1^0 \sigma_2(n) = \frac{1}{6}\varphi(n)[2n^2 + (-1)^k p_1 \cdots p_k]$$

$$2^0 \sigma_3(n) = \frac{1}{4} n^3 \varphi(n) \left[1 + \frac{(-1)^k p_1 \cdots p_k}{n^2}\right]$$

$1^0$ We take in (10) and (11): $\underline{f(x) = x^2}$. Then

$$F_d(n) = \sum_{k=1}^{n|d} (kd)^2 = d^2 \cdot \sum_{k=1}^{n|d} k^2 = d^2 \cdot S_2\left(\frac{n}{d}\right)$$

We have

$$S_2(n) = 1^2 + 2^2 + \cdots + n^2 = \frac{1}{6}n(n+1)(2n+1)$$

so that

$$F_d(n) = d_2 \cdot \frac{1}{6}\frac{n}{d}\left(\frac{n}{d}+1\right)\left(\frac{2n}{d}+1\right)$$
$$= \frac{1}{6}\frac{1}{d}(2n^3 + 3n^2 d + nd)$$

Therefore the formula (11) yields

$$\sigma_2(n) = \frac{1}{6}\left\{2n^3 . \sum_{d|n}\frac{\mu(d)}{d} + 3n^2\sum_{d|n}\mu(d) + n\sum_{d|n}d\mu(d)\right\}$$

Now from Th.2, (4) and (5), (See. P. 31)

$$\sum_{d|n}\frac{\mu(d)}{d} = \left(1-\frac{1}{p_1}\right)\cdots\left(1-\frac{1}{p_k}\right) = \frac{\varphi(n)}{n}$$

$$\sum_{d|n}\mu(d) = 0$$

$$\sum_{d|n}d\mu(d) = (1-p_1)\ldots(1-p_k)$$

$$= (-p_1)\left(1-\frac{1}{p_1}\right)\cdots(-p_k)\left(1-\frac{1}{p_k}\right)$$

$$= (-1)^k p_1\cdots p_k\frac{\varphi(n)}{n}$$

Consequently

$$\sigma_2(n) = \frac{1}{6}\left[2n^2\varphi(n) + (-1)^k p_1\cdots p_k\varphi(n)\right]$$

$2^0$ We take in (10) and (11) $f(x) = x^3$. Then

$$F_d(n) = \sum_{k=1}^{n|d}(kd)^3 = d^3 \cdot \sum_{k=1}^{n|d}k^3 = d^3 \cdot S_3\left(\frac{n}{d}\right)$$

Since $S_3(n) = 1^3 + \cdots + n^3 = \frac{1}{4}n^2(n+1)^2$ we have

$$F_d(n) = d^3 \cdot \frac{1}{4}\frac{n^2}{d^2}\cdot\frac{(n+d)^2}{d^2} = \frac{n^2}{4}\left(\frac{n^2}{d} + 2n + d\right)$$

Consequently the formula (11) yields

$$\sigma_3(n) = \frac{n^2}{4}\left\{n^2 \cdot \sum_{d|n}\frac{\mu(d)}{d} + 2n\sum_{d|n}\mu(d) + \sum_{d|n}d\mu(d)\right\}$$

$$= \frac{n^2}{4}\left\{n^2 \cdot \frac{\varphi(n)}{n} + (-1)^k p_1 \cdots p_k \frac{\varphi(n)}{n}\right\}$$

thus

$$\sigma_3(n) = \frac{n^3\varphi(n)}{4}\left\{1 + \frac{(-1)^k p_1 \cdots p_k}{n^2}\right\}$$

We shall formulate Theorem 4 in a slightly different form. Consider a real or complex valued function $f(x)$ defined an $0 \leq x \leq 1$, and consider the set $S = \{\frac{x}{n},\ \ x = 1, \ldots, n\}$. This essentially amounts to replacing $f(x)$ by $f(\frac{x}{n})$ in (10) and (11)

Let us set

$$F(n) = \sum_{k=1}^{n} f\left(\frac{k}{n}\right) \text{ and } F^*(n) = \sum_{\substack{1 \leq x \leq n \\ (x,n)=1}} f\left(\frac{x}{n}\right) \tag{12}$$

Then (11) becomes

$$F^*(n) = \sum_{d|n} \mu(d) F_d(n) \text{ where } F_d(n) = \sum_{k=1}^{n|d} f\left(\frac{kd}{n}\right)$$

From (12)

$$F_d(n) = F\left(\frac{n}{d}\right)$$

So that we now have

$$F^*(n) = \sum_{d|n} \mu(d) F\left(\frac{n}{d}\right)$$

In this manner:

**Theorem 5.**

*If* $F(n) = \sum_{k=1}^{n} f\left(\frac{k}{n}\right)$ *and* $F^*(n) = \sum_{\substack{1 \leq x \leq n \\ (x,n)=1}} f\left(\frac{x}{n}\right)$, *then*

$$\boxed{F^*(n) = \sum_{d|n} \mu(d) F\left(\frac{n}{d}\right)}$$

**Theorem 6. [Möbius Inversion formula]**

$$(A)\quad \boxed{\sum_{d|m} f(d) = g(m) \Longleftrightarrow \sum_{d|m} \mu\left(\frac{m}{d}\right) g(d) = f(m)} \quad (B)$$

*Preliminary Remark*

It is easily verified that, when $t|d|m$, for each fixed $t|m$:

(i) $\frac{m}{d}$ runs through all the divisors of $\frac{m}{t}$ and

(ii) $\dfrac{d}{t}$ runs through all the divisors of $\dfrac{m}{t}$

when $d$ runs through all the divisors of $m$ (with $t|d$); we observe that

$$\frac{m}{t} = \frac{m}{d} \cdot \frac{d}{t}$$

The fundamental property of the Mobius function (See 5 P. 30) can be written as follows:

$$\sum_{d|m} \mu(d) = \varepsilon(m) \text{ where } \varepsilon(m) = \begin{vmatrix} 1 \text{ for } m = 1 \\ 0 \text{ for } m \neq 1 \end{vmatrix}$$

**Proof of Möbius formula**

$(A) \Rightarrow (B)$: Assume $(A)$, and substitute for $g(d)$ on the left side of $(B)$. Then

$$g(d) = \sum_{t|d} f(t)$$

so that the left side of $(B)$ becomes

$$\sum_{d|m} \mu\left(\frac{m}{d}\right)\left\{\sum_{t|d} f(t)\right\} = \sum_{t|d|m} f(t)\mu\left(\frac{m}{d}\right) =$$

$$= \sum_{t|m} f(t)\left\{\sum_{\frac{m}{d}|\frac{m}{t}} \mu\left(\frac{m}{d}\right)\right\} = \sum_{t|m} f(t)\varepsilon\left(\frac{m}{t}\right) = f(m)$$

$(B) \Rightarrow (A)$: Assume $(B)$ and substitute for $f(d)$ on the left side of (A). Then

$$\sum_{t|d} \mu\left(\frac{d}{t}\right) g(t) = f(d)$$

so that the left side of $(A)$ becomes

$$\sum_{d|m}\left\{\sum_{t|d} \mu\left(\frac{d}{t}\right) g(t)\right\} = \sum_{t|d|m} \mu\left(\frac{d}{t}\right) g(t)$$

$$= \sum_{t|m} g(t)\left\{\sum_{\frac{d}{t}|\frac{m}{t}} \mu\left(\frac{d}{t}\right)\right\} = \sum_{t|m} g(t)\varepsilon\left(\frac{m}{t}\right) = g(m)$$

This completes the proof of the Möbius Inversion formula ■

We shall make an application of the Mobius Formula. Consider the set of integers $S = \{1, 2, \ldots, m\}$. Fixing the modulus $m$ we partition the set $S$ into subsets $A_d$ as follows:

For every divisor $d$ of $m$, set $x \in A_d$
if and only if $(x, m) = d$

$$S = 1, \ldots, m = \bigsqcup_{d|m} A_d \text{ (disjoint union)}$$

$$x \in A_d \Longleftrightarrow (x, m) = d$$

If $n_d$ is the number of elements in $A_d$, then

$$m = \sum_{d|m} n_d$$

Since $(x, m) = d$ if and only if $(\frac{x}{d}, \frac{m}{d}) = 1$ the number of integers $x$ with $1 \leq x \leq m$ and $(x, m) = d$ is the same as the number of integers $\leq \frac{m}{d}$ and coprime with $\frac{m}{d}$, i.e.

$$n_d = \varphi\left(\frac{m}{d}\right)$$

Consequently we obtain

(13) $$\boxed{\sum_{d|m} \varphi\left(\frac{m}{d}\right) = m} \quad \text{or} \quad \boxed{\sum_{d|m} \varphi(d) = m}$$

Hence we take in Theorem 6:

$$f(m) = \varphi(m) \quad \text{and} \quad g(m) = m$$

and obtain

$$\sum_{d|m} \mu\left(\frac{m}{d}\right) \cdot d = \varphi(m)$$

When $d$ runs through all divisors of $m$, so does $\frac{m}{d}$; so we can replace in $\sum_{d|m}$ the divisors $d$ by $\frac{m}{d}$, thus

$$\sum_{d|m} \mu(d) \frac{m}{d} = \varphi(m)$$

or

(14) $$\boxed{\sum_{d|m} \frac{\mu(d)}{d} = \frac{\varphi(m)}{m}}$$

[Compare (13) and (14) with the results obtained in Theorem 4, p. 34, Example 2, p. 34 and Theorem 5, p. 36 ].

*Remark*

Given any real or complex valued function $f$ defined on the rationals we often have to consider sums of the following type

$$(15)\qquad \boxed{F(n) = \sum_{k=1}^{n} f\left(\frac{k}{n}\right) \text{ and } F^*(n) = \sum_{\substack{x=1\\(x,n)=1}}^{n} f\left(\frac{x}{n}\right)}$$

If $S$ is the index set $S = \{1, 2, \ldots n\}$ we partition it into subsets $A_d$, $(d|n)$, as follows:

$$A_d = \{x \in S,\ (x,n) = d\}$$
$$S = \bigsqcup_{d|n} A_d \text{ (disjoint union)}$$

$$(15')\qquad \sum_{k\in S} h(k) = \sum_{d|n} \sum_{k\in A_d} h(k)$$

for any function $h$ defined on $S$. We have for $d|n$:

$$k \in A_d \Longleftrightarrow (k,n) = d \Longleftrightarrow \left(\frac{k}{d}, \frac{n}{d}\right) = 1$$

Put $\dfrac{k}{d} = k'$, $\dfrac{n}{d} = n'$, then we have

$$\frac{k}{n} = \frac{k'}{n'},\quad 1 \le k' \le n' \text{ and } (k'n') = 1$$

Hence

$$F(n) = \sum_{k=1}^{n} f\left(\frac{k}{n}\right) = \sum_{d|n}\left\{\sum_{\substack{k'=1\\(k',n')=1}}^{n'} f\left(\frac{k'}{n'}\right)\right\}$$
$$= \sum_{d|n} F^*(n') = \sum_{d|n} F^*\left(\frac{n}{d}\right)$$

Consequently (see (15) and (15')) the two sums $F(n)$ and $F'(n)$ are related by the identity

$$\boxed{F(n) = \sum_{d|n} F^*\left(\frac{n}{d}\right)} \qquad (16)$$

We observe that in the sum $F(n)$ the index $k$ runs through $1, \ldots, n$ whereas in the sum $F^*(n)$ the index $x(1 \le x \le n)$ is restricted by the condition $(x,n) = 1$.

The Mobius Inversion formula can now be applied to (16) to yield

$$(17)\qquad \boxed{F^*(n) = \sum_{d/n} \mu^{(\delta)}\ F\left(\frac{n}{d}\right)}$$

[Compare Th. 4, p. 34 and Th. 5, p. 36].

## § 7. CONGRUENCES

We fix a natural number m and call it a *modulus*. Two integers $a$ and $b$ are said to be *congruent mod. m* if they leave the same remainder when divided by $m$. We write then

$$a \equiv b(\operatorname{mod} m)$$

**Theorem 1.**

$a \equiv b(\operatorname{mod} m)$ *if and only if* $m|(a-b)$

**Proof:** If $a \equiv b(\operatorname{mod} m)$ we have, by definition

$$a = mq + r, \quad b = mq' + r \qquad (0 \leq r \leq m)$$

whence

$$a - b = m(q - q'), \quad a = b + mt \text{ with } t = q - q'$$

$$a - b = m.t. \quad m|(a - b)$$

Conversely, let $m|a-b$, $a = b + tm$. Divide $b$ by the modulus $m$:

$$b = mq' + r, \quad 0 \leq r \leq m$$

Then $a = (mq' + r) + tm = m(q' + t) + r$, i.e. $a$ leaves the same remainder when divided by $m$.

**Theorem 2.**

*(All congruences are with respect to one and the same modulus $m$)*

I) $a \equiv a$, *if* $a \equiv b$, *then* $b \equiv a$; $a \equiv b$ *and* $b \equiv c$, *imply* $a \equiv c$

II) $a \equiv a'$, *and* $b \equiv b'$ *imply* $a + b \equiv a' + b'$, $ka \equiv ka'$, $ab \equiv a'b'$

III) *If $f$ is a polynomial with integer coefficients, then*

$$a \equiv b \text{ implies } f(a) \equiv f(b)$$

IV) *(Cancellation Law) The common factor $k$ in the congruence* $ka \equiv kb(\operatorname{mod} m)$ *can be cancelled* **provided** $(k, m) = 1$

V) *If* $a \equiv b(\operatorname{mod} m)$, *then, for any* $k \neq 0$, $\quad ak \equiv bk(\operatorname{mod} mk)$

VI) *If* $a \equiv b(\operatorname{mod} m)$ *and $\delta$ divides $a$, $\;b$ and $m$ then*

$$\frac{a}{\delta} \equiv \frac{b}{\delta}\left(\operatorname{mod} \frac{m}{\delta}\right)$$

**Proof:** All the results stated are direct consequences of the definitions. We prove the last three statements.

If $ka \equiv kb(\operatorname{mod} m)$, then $m|k(a-b)$; If $(k, m) = 1$ this implies $m|a-b$, i.e. $a \equiv b$. This proves IV.

If $a \equiv b(\bmod m)$, then $a - b = t \cdot m$ so that $ak - bk = t \cdot mk$, therefore $ak \equiv bk(\bmod mk)$.

We prove VI. When $\delta$ divides $a$, $b$ and $m$, put

$$a = \delta a', \quad b = \delta b', \quad m = \delta m'$$

If $a \equiv b(\bmod m)$ we can write $a = b + tm$, so

$$\delta a' = \delta b' + t \cdot \delta m' \text{ i.e. } a' = b' + tm'$$

which means $a' \equiv b'(\bmod m)$. This proves VI ■

*Remark*

In a congruence terms on either side which are $\equiv 0(\bmod m)$ may be omitted.

**Theorem 3.**

*Two numbers congruent mod m have the same greatest common divisor with the modulus*

$$(1) \qquad \boxed{a \equiv b(\bmod m) \Rightarrow (a, m) = (b, m)}$$

In particular

$$(2) \qquad \boxed{a \equiv b(\bmod m) \text{ and } (a, m) = 1 \Longrightarrow (b, m) = 1}$$

In fact, we have $a = b + tm$ whence follows that every common divisor of $a$ and $m$ must also divide $b$. In particular, the greatest common divisors must coincide:

*Definition 1*

If $Z = \{0, \pm 1, \pm 2, \ldots\}$ is the set of all integers, and the modulus $m$ is fixed , let $C_k$ denote the set of all integers in $Z$ which are $\equiv k(\bmod\ m)$; every integer in $C_k$ yields the remainder $k$ when divided by $m$:

$$k = 0, 1, 2, \ldots, m - 1$$

Since these exhaust all possible remainders every element in $Z$ lies in one and only one of the classes

$$C_0, C_1, C_2, \ldots, C_{m-1}$$

We have seen (Th. 2, I) that the congruence is an equivalence relation:

$$a \equiv a \text{ (reflexive)}$$

$$a \equiv b \Longleftrightarrow b \equiv a \text{ (symmetric)}$$

$$a \equiv b \text{ and } b \equiv c \Longleftrightarrow a \equiv c \text{ (transitive)}$$

Accordingly the set $Z$ gets partitioned into equivalence classes, namely: $C_0, C_1, C_2, \ldots, C_{m-1}$.

(3) $$Z = C_0 \dot{+} C_1 \dot{+} C_2 \dot{+} \cdots \dot{+} C_{m-1} \quad \text{(disjoint union)}$$

Any system of $m$ numbers, ($m$ is the modulus), taken one from each class is called a *complete system of residues mod. m.* Thus

(i) every integer in $Z$ lies in one and only one of the classes $C_0, C_1, \ldots C_{m-1}$

(ii) Any two numbers in one and the same clas are always congruent mod. $m$

(iii) If $a$ and $b$ are in different classes then $a \not\equiv b (\text{mod } m)$

**Example 1.** Let $m = 6$. Take

$$R = \{10, 65, 26, 12, 13, 33\}$$

when divided by 6 these yield the remainders

$$\{4, 5, 2, 0, 1, 3\}$$

These exhaust all the possible remainders mod. 6: namely, 0, 1, 2, 3, 4, 5. Hence $R$ forms a complete system of residues (mod. 6)

We have seen in Th. 3 that two numbers congruent mod. $m$ have with $m$ the same greatest common divisor. In (3)

$$C_k = \{a \in Z, \quad a \equiv k(\text{mod}\,.m)\}$$

so that

(4)
$1^0 (x, m) = k$ for $x \in C_k$ and $k/m (k = 1, 2, \ldots, m - 1)$

$2^0 (x, m) = m$ for $x \in C_0 = C_m$

i.e.$m|x$ or $x \equiv 0(\text{mod}\,.m)$ thus also

$$x \equiv m(mod.m)$$

$3^0$ if one element in $C_k$ is coprime with $n$ then every element in $C_k$ must be coprime with $m$. We call such classes *prime residue classes (mod. m).*

Accordingly the prime residue classes (mod. $m$) are given by

$$C_{r_1}, C_{r_2}, \ldots, C_{r_h}$$

where each one of the $r_1, r_2, \ldots, r_h$ must be such that

$$\{r_1, r_2, \ldots, r_h\} \subset \{1, 2, \ldots, m\}$$

$$r_1 \neq r_2 \neq \ldots \neq r_h$$

$$(r_k, m) = 1 \text{ for } k = 1, \ldots h$$

This implies that $h = \varphi(m)$ (the Euler $\varphi$ function), thus:

(5) **The number of prime residue classes mod.$m = \varphi(m)$**

Any system of $h = \varphi(m)$ numbers, one taken from each of the classes $C_{r_1}, \ldots, C_{r_h}$ is said to form a *reduced system of residues (mod. m)*

**Example 1 (continued):** $m = 6$

The numbers in $\{1, 2, \ldots, 6\}$ that are coprime with 6 are

$$\{1, 5\}, \qquad h = \varphi(6) = 2$$

Hence any two numbers $\{r_1, r_2\}$ such that $r_1 \equiv 1(\text{mod } 6)$ and $r_2 \equiv 5(\text{mod } 6)$ will form a reduced system mod. 6

Consequently:

A system $S$ of numbers will form a reduced residue system (mod $m$) if and only if the following conditions are satisfied:

(6)
$1^0 S$ must consist of $h$ numbers, $(h = \varphi(m))$;
$2^0$ every element in $S$ is co-prime with $m$;
$3^0$ any two integers $a$ and $b$ in $S$ are incongruent mod.$m : a \not\equiv b(mod.m)$

**Theorem 4.**

$\text{I}^0$ (EULER) $a^{\varphi(m)} \equiv 1(\text{mod } m)$ provided $(a, m) = 1$

$\text{II}^0$ (FERMAT) If $a$ is any natural number, and $p$ is any prime
$a^{p-1} \equiv 1(\text{mod } p)$ provided $p \dagger a$ and
$a^p \equiv a(\text{mod } p)$ for all $a$ †.

Fermat announced the theorem in 1640 without proof. In 1760 Euler gave the first proof of his generalization of Fermat's theorem.

**Proof:** Denote by $r_1, \ldots, r_h$ all the distinct $h$ numbers, $h = \varphi(m)$, among $1, 2, \ldots, m$ that are coprime with $m$:

$$\{r_1, \ldots, r_h\} \subset \{1, \ldots, m\}, \quad h = \varphi(m)$$
$$r_1 \neq \ldots \neq r_h$$

Consider the system $S = \{ar_1, \ldots, ar_h\}$

*Claim:* $S$ satisfies the three conditions in (6) for a reduced residue system mod $m$.

In fact $S$ satisfies the condition $1^0$ since it consists of $h$ numbers. Since $(a, m) = 1$ and $(r_k, m) = 1$ $(k = 1, \ldots, h)$ we have

$$(ar_k, m) = 1 \ (k = 1, \ldots, h)$$

i.e. $S$ satisfies the condition $2^0$. In order to verify the condition $3^0$ we have to prove

$$ar_i \not\equiv ar_j (\mathrm{mod}\, m) \text{ for } i \neq j$$

This is certainly true, since, otherwise

$$ar_i \equiv ar_j (i \neq j) \text{ implies } r_i \equiv r_j (\mathrm{mod}\, m)$$

because the common factor $a$ can be cancelled, being coprime with the modulus $m$.

Accordingly each $ar_i$ is congruent to one and only one among $r_1, \ldots, r_h$, thus

(7) $$ar_i \equiv r'_1, \ldots, ar_h \equiv r'h \quad \{r'_1 \ldots r_h\} = \{r_1, \ldots r_h\}$$

wherein $r'_1, \ldots, r'_h$ are the same elements, $r_1, \ldots, r_h$ in some other order. But then their products coincide:

(8) $$r'_1 \ldots, r'_h = r_1 \ldots r_h$$

From (7) and (8)

$$(ar_1)\ (ar_2) \ldots (ar_h) \equiv r'_1 r'_2 \ldots r'_h = r_1, r_2 \cdots r_h \ (\mathrm{mod}\, m)$$

Put

$$\varrho = r_1 \ldots r_h$$

Since each $r_i$ is coprime with $m$ we must have $(\varrho, m) = 1$

Hence, from (9),

$$a^h \cdot \varrho \equiv \varrho (\mathrm{mod}\, m)$$

Herein, since $(\varrho, m) = 1$, $\varrho$ can be cancelled; we get

$$a^h \equiv 1 (\mathrm{mod}\, m), \quad (h = \varphi(m))$$

This proves the theorem $\mathrm{I}^0$ of Euler

Fermat's theorem $\mathrm{II}^0$ at once follows since:

$$m = p \text{ (prime) implies } \varphi(m) = p - 1$$

$$(a, p) = 1 \text{ means } p \dagger a$$

Finally

(10) $$a^p \equiv a (\mathrm{mod}\, p) \cdot (p \text{ prime }), \text{ for all } a$$

In fact, if $(a, p) = 1$, i.e., $p \dagger a$, we have as was just proved $a^{p-1} \equiv 1 (\mathrm{mod}\ p)$ so that

$$a \cdot a^{p-1} \equiv a \cdot 1, \quad a^p \equiv a (\mathrm{mod}\, p)$$

On the other hand, if $p|a$,

$$a \equiv 0(\bmod p), \text{ so } a^p \equiv 0(\bmod p), \text{ and } a^p \equiv a(\bmod p)$$

This completes the proof of Euler's and Fermat's theorems.

*Remarks*

We have seen that if $r$ is the remainder left by $a$ when divided by $m$

$$a = q.m + r, \quad 0 \leq r < m$$

one has

$$a \equiv r(\bmod m)$$

$r$ is called the *least positive residue* of $a \bmod .m$. It is *uniquely* determined by the conditions

$$a \equiv r(\bmod m) \textbf{ and } 0 \leq r < m$$

Now (Th. 2, See p. 40) congruences can be added, multiplied and raised to arbitrary powers $k(k = 1, 2, \ldots)$ i.e. The congruences

$$a \equiv a', \quad b \equiv b', \quad c \equiv c' \quad (\bmod m)$$

imply

$$a + b \equiv a' + b', \quad a - b \equiv a' - b', \quad ka \equiv ka', \quad abc \equiv a'b'c',$$

$$a^n \equiv a'^n \quad (n = 1, 2, \ldots)$$

It is useful to remember that terms in a congruence which are $\equiv 0(\bmod\ m)$ can be neglected:

$$a + b + c \equiv a' + b' \quad (\bmod m)$$

$$c \equiv 0 \text{ and } a' \equiv 0 \quad (\bmod m)$$

imply

$$a + b \equiv b' \quad (\bmod m)$$

The flexibility of the congruence notion, introduced by GAUSS, will be clear from the following examples.

**Example 2.** Find the remainder obtained by dividing

$$[12371^{56} + 34]^{38} \text{ by } 111$$

Put $m = 111 = 3.37 = 3p$, $p = 37$ (prime), and let $a = 12371$. It is verified that $a$ is neither divisible by 3 nor by 37, thus

$$(a, 3) = 1, \quad (a, 37) = 1, \quad (a, m) = 1$$

Since $(a, p) = 1$, by Fermat's theorem $a^{p-1} \equiv 1(\text{mod } p)$ i.e

(11) $$a^{36} \equiv 1(\text{mod}\, 37)$$

Denoting the number in question by $N$,

$$N = A^{28} \text{ where } A = a^{56} + 34 \text{ and } a = 12371$$

Dividing a by 37 we obtain the remainder 13, thus

$$\begin{array}{r|r|l} 37 & 12371 & 334 \\ & 111 & \\ \hline & 127 & \\ & 111 & \\ \hline & 161 & \\ & 148 & \\ \hline & 13 & \end{array}$$

(12) $a \equiv 13(\text{mod}\, 37)$

By successive squaring

$$a^2 \equiv 169 \equiv 21(\text{mod}\, 37)$$

(12′) $a^4 \equiv 21^2 \equiv 441 \equiv 34 \equiv -3(\text{mod}\, 37)$

$$a^8 \equiv, a^{16} \equiv 81 \equiv 7(\text{mod}\, 37)$$

$$a^{20} \equiv a^4.a^{16} \equiv (-3).(7) \equiv (-21) \equiv 16,$$

(13) $a^{20} \equiv 16(\text{mod}\, 37)$

Therefore

$$A = a^{56} + 34 \equiv a^{36} \cdot a^{20} + 34 \equiv 1.16 + 34, \qquad \textit{See(11) and (13)}$$

$$\equiv 50 \equiv 13(\text{mod}\, 37)$$

We get

$$N = A^{28} \equiv 13^{28}(\text{mod}\, 37)$$

Since from (12) and (12') and (13)

$$a \equiv 13, \quad a^8 \equiv \dot{9}, \quad a^{20} \equiv 16$$

We

$$N \equiv 13^{28} \equiv a^{28} \equiv a^8 \cdot a^{20} \equiv 9 \cdot 16 \equiv 144 \equiv 33$$

i.e.

(14) $$N \equiv 33(\text{mod}\, 37)$$

We now look at the other prime factor 3 of $111 = 3p$, $(p = 37)$. Dividing a by 3 we get the remainder 2:

$$a \equiv 2(\text{mod}\, 3), \ a^2 \equiv 4 \equiv 1(\text{mod}\, 3)$$

$$a^{56} \equiv (a^2)^{28} \equiv 1(\text{mod}\, 3)$$

$$A = a^{56} + 34 \equiv 1 + 34 = 35 \equiv -1(\text{mod}\, 3)$$

$$A^{28} \equiv (-1)^{28} \equiv 1(\text{mod}\, 3)$$

(15) $$N \equiv 1(\text{mod}\, 3)$$

From (14) we can express $N$ in the form $N = 33 + 37t$.
Substituting in (15)

$$33 + 37t \equiv 1(\bmod 3) \text{ or } 37t \equiv 1(\bmod 3)$$

Observe that 33 can be neglected since $33 \equiv 0(\bmod\ 3)$.
Since $37 \equiv 1\ (\bmod\ 3)$ we now have $t \equiv 1(\bmod\ 3)$. Therefore

$$N = 33 + 37t \equiv 33 + 37 = 70(\bmod 3)$$

while, from (14),

$$N \equiv 33 \equiv 70(\bmod 37)$$

In this manner we have

$$N \equiv 70(\bmod 37) \text{ and } N \equiv 70(\bmod 3)$$

Now

$$x \equiv y(\bmod m_1)$$

$$x \equiv y(\bmod m_2) \text{ and } (m_1, m_2) = 1 \text{ imply } x \equiv y(\bmod m_1 m_2)$$

Consequently

$$N \equiv 70(\bmod 111)$$

Since the right member in this congruence is non-negative and less than the modulus 111, it follows that $N$ leaves the remainder 70 when divided by $m$.

In connection with Fermat's theorem: $a^{p-1} \equiv 1(\bmod p)$, $p$ prime, $a \not\equiv 0(\bmod p)$ Abel (1828) raised the question:

Do there exist integers $a$ and primes $p$, $a \not\equiv 0(\bmod\ p)$, such that

$$a^{p-1} \equiv 1(\bmod p^2) \tag{16}$$

**Example 3.** $3^{10} \equiv 1(\bmod 11^2)$
We have $3^5 = 243 = 1 + 121.2 \equiv 1(\bmod 11^2)$ so that

$$3^{10} \equiv 1(\bmod 11^2)$$

Again $9^{10} = 3^{20} = (3^{10})^2 \equiv 1(\bmod 11^2)$ so that

$$9^{10} \equiv 1(\bmod 11^2)$$

So, for $p = 11$, we can take $a = 3$ or 9.

Jacobi verified that for primes $p \leq 37$, the congruence (16) is possible only for the following values:

| $a$ | 3 | 9 | 14 | 18 |
|---|---|---|---|---|
| $p$ | 11 | 11 | 29 | 37 |

**Example 4.** Show that for the prime $p = 1093$ the congruence (16) is satisfied by $a = 2$:

$$\boxed{2^{1092} \equiv 1 (\mathrm{mod}\, 1093^2)}$$

Put $p = 1093$. We observe that

$$3^7 = 2187 = 2p + 1 \text{ and } 2^{14} = 16384 = 15p - 11 \tag{17}$$

Using these relations we prove

$$\text{I}^\circ \qquad 3^{14} \equiv 4p + 1 (\mathrm{mod}\, p^2)$$

$$\text{II}^\circ \quad 3^{14} \cdot 2^{182} \equiv -4p - 1 (\mathrm{mod}\, p^2)$$

In fact we obtain $\text{I}^0$ immediately from (17):

$$3^{14} = (2p + 1)^2 \equiv 1 + 4p (\mathrm{mod}\, p^2)$$

Since $2^{14} = 15p - 11$,

$$2^{28} = (15p - 11)^2 \equiv 121 - 330p (\mathrm{mod}\, p^2)$$

so that

$$3^2 \cdot 2^{28} \equiv 1089 - 2970p \equiv -4 + p - 2970p \equiv -4 - 2969p$$
$$\equiv -4 - (3p - 310)p \equiv -4 + 310p (\mathrm{mod}\, p^2)$$

i.e.

$$3^2 . 2^{28} \equiv -4 + 310p (\mathrm{mod}\, p^2)$$

Multiply both sides by 7 to get

$$3^2 \cdot 2^{28} \cdot 7 \equiv -28 + 2170p \equiv -28 + (2p - 16)p$$
$$\equiv -28 - 16p (\mathrm{mod}\, p^2)$$

i.e

$$3^2 \cdot 2^{28} \cdot 7 \equiv (-4p - 7) \cdot 2^2 (\mathrm{mod}\, p^2)$$

Cancelling the common factor $2^2$, we get

$$3^2 \cdot 2^{26} \cdot 7 \equiv -(7 + 4p) (\mathrm{mod}\, p^2)$$

Raising this congruence to the power 7 we have

$$3^{14} \cdot 2^{182} \cdot 7^7 \equiv -(7 + 4p)^7$$
$$\equiv -[7^7 + 7.7^6 .4p] (\mathrm{mod}\, p^2),$$

using the binomial expansion for $(7 + 4p)^7$. Hence

$$3^{14} \cdot 2^{182} \cdot 7^7 \equiv -(1 + 4p) \cdot 7^7 (\mathrm{mod}\, p^2)$$

Cancelling the common factor $7^7$ we have

$$3^{14} \cdot 2^{182} \equiv -(1+4p)(\text{mod}\, p^2)$$

But from I$^0$

$$3^{14} \equiv (1+4p)(\text{mod}\, p^2)$$

Cancelling the factor $3^{14}$ we get

$$\boxed{2^{182} \equiv -1(\text{mod}\, p^2)}$$

Whence follows

$$2^{1092} = (2^{182})^6 \equiv (-1)^6 \equiv 1(\text{mod}\, p^2)$$

which proves the required result.

**Example 5.** Show that $\boxed{18^{36} \equiv 1(\text{mod}\, 37^2)}$

$p = 37$ is a prime. We observe that $6^2 = p - 1$ so that

$$6^{36} = (p-1)^{18} = 1 - 18p + \binom{18}{2} p^2 - \cdots \equiv 1 - 18p(\text{mod}\, p^2)$$

i.e.

(18) $$6^{36} \equiv 1 - 18p(\text{mod}\, p^2)$$

Next we have $3^3 = p - 10$ so tht

$$3^6 = (p-10)^2 \equiv 100 - 20p \equiv (3p-11) - 20p \equiv -(11+17p)$$

$$3^{12} \equiv (11+17p)^2 \equiv 121 + 2 \cdot 11 \cdot 17p(\text{mod}\, p^2)$$

$$\equiv (3p+10) + (10p+4)p \equiv 7p + 10(\text{mod}\, p^2)$$

$$3^{36} \equiv (7p+10)^3 \equiv 10^3 + 3.10^2.7p(\text{mod}\, p^2)$$

Now $10^2 = 3p - 11$, $10^3 = 30p - 110 = (p-7)p - (3p-1) \equiv 1 - 10p(\text{mod}\, p^2)$ and

$$3.10^2.7 = 21(3p-11) = 63p - (6p+9) = 57p - 9$$

$$3.10^2.7p = 57p^2 - 9p \equiv -9p(\text{mod}\, p^2)$$

Hence

$$3^{36} \equiv (1-10p) + (-9p) \equiv 1 - 19p$$

Thus we have

(19) $$3^{36} \equiv 1 - 19p(\text{mod}\, p^2)$$

Comparing (18) and (19) we have:

$$18^{36} \equiv 6^{36} \cdot 3^{36} \equiv (1-18p)(1-19p)$$

$$\equiv 1 - 37p(\text{mod}\, p^2)$$

$$\equiv 1(\text{mod}\, p^2) \text{ since } p = 37 \blacksquare$$

Given a polynomial $f(x) = a_0x^n + a_1x^{n-1} + \cdots + a_n$, where $a_0, a_1, \ldots, a_n$ are integers, we consider the problem of solving the congruence

$$f(x) \equiv 0 (\text{mod}\, m) \tag{20}$$

If the highest coefficient $a_0$ is such that

$$a_0 \not\equiv 0 (\text{mod}\, m)$$

then the congruence is of degree $n$.

If $x_0$ is a solution of (20), i.e. $f(x_0) \equiv 0 (\text{mod}\, m)$ and $x_1 \equiv x_0 (\text{mod}\, m)$, we have by the basic properties of congruences (Th. 2, III, p. 40).

$$f(x_1) \equiv f(x_0) \equiv 0 (\text{mod}\, m)$$

i.e. $x_1$ is again a solution of (20). Thus if $x_0$ is a solution of the congruence all the integers in the residue class of $x_0 (\text{mod}\, m)$ are also solutions. Accordingly we agree to stipulate that $x_0$ and $x_1$ are *distinct* solutions of (20) if and only if they are incongruent mod $m$:

$$f(x_0) \equiv 0, \quad f(x_1) \equiv 0, \text{ and } x_0 \not\equiv x_1 (\text{mod}\, m)$$

So we need only search for the solutions of (20) in any complete system of residues mod $m$, e.g. the least positive system of residues mod $m$

$$x = 0, 1, 2, \ldots, m - 1$$

**Example 6**

To solve: $x^4 \equiv 1 (\text{mod}\, 5)$

For $x = 1$, $x^4 = 1 \equiv 1$; for $x = 2$, $x^4 = 2^4 = 16 \equiv 1$; for $x = 3x^4 = 3^4 = 81 \equiv 1$, for $x = 4$, $x^4 = 4^4 = 256 \equiv 1$. Consequently the congruence has the four distinct solutions

$$x = 1, 2, 3, 4$$

i.e. as many as the degree of the congruence.

The congruence

$$x^5 \equiv x (\text{mod}\, 5)$$

has a complete system of residues mod $5 : \{0, 1, 2, 3, 4\}$ as its solutions

This is as it should be since, by Fermat's theorem,

$$x^p \equiv x (\text{mod}\, p), \quad p \text{ prime}$$

for every $x$.

**Example 7**

To solve: $x^5 + x + 1 \equiv 0 \pmod 7$
put $f(x) = x^5 + x + 1$. We have

$$1^5 \equiv 1, \quad 2^5 \equiv 4, \quad 3^5 \equiv 3^3.3^2 \equiv (-1)2 \equiv 5, \quad 4^5 \equiv 4^3 \cdot 4^2 \equiv 1 \cdot 2 \equiv 2$$

$$5^5 \equiv 5^3 \cdot 5^2 \equiv 6 \cdot 4 \equiv 3, \quad 6^5 \equiv 6^3 \cdot 6^2 \equiv 6 \cdot 1 \equiv 6$$

so that

$$f(1) \equiv 3, \quad f(2) \equiv 0, \quad f(3) \equiv 2, \quad f(4) \equiv 0, \quad f(5) \equiv 2, \quad f(6) \equiv 6$$

Hence we get the two distinct solutions $x_1 = 2$, $x_2 = 4$:

$$x_1 \equiv 2 \text{ and } x_2 \equiv 4 \pmod 7$$

**Example 8.** To solve $8x \equiv 12 \pmod{52}$
Since 4 is a common factor of 8,12 and 52, we can cancel it and obtain the congruence

$$2x \equiv 3 \pmod{13}$$

This is satisfied by $x_0 = 8$. We have

$$x \equiv x_0 \pmod{13}, \quad x = 8 + 13t$$

Herein for $t = 0, 1, 2, 3$ we get

$$8, \quad 8 + 13, \quad 8 + 26, \quad 8 + 39$$

as again solutions of $2x \equiv 3 \pmod{13}$, but they are all congruent to each other mod 13. However they are distinct solutions of the original congruence $8x \equiv 12 \pmod{52}$ because the numbers 8,21,34,47 are mutually incongruent (mod 52);

**Theorem 5.**

I⁰ *The linear congruence (or congruence of the first degree)*

(21) $\boxed{ax + b \equiv 0 \pmod m}$ with $a \not\equiv 0 \pmod m$

*has a unique solution, provided* $(a, m) = 1$

II⁰ *Let* $g = (a, m) > 1$. *Then if* $g$ *does not divide* $b$ *there is no solution*

III⁰ *Assume* $g > 1$ *and* $g|b$. *Put*

$$\frac{a}{g} = a_0, \quad \frac{b}{g} = b_0, \quad \frac{m}{g} = m_0; \text{ then } (a_0, m_0) = 1$$

*Then, by I°, the congruence*

$$a_0x + b_0 \equiv 0(\text{mod}\, m_0)$$

*has a unique solution $x_1$ such that $0 \leq x_1 \leq m_0$. Then all the distinct* (mod $m$) *solutions of the congruence (21) are given by*

(22) $$x_1, x_1 + m_0, x_1 + 2m_0, \ldots, x_1 + (g-1)m_0$$

The congruence (21) has $g$ distinct solutions (i.e. incongruent mod $m$)

*Proof:* We know [See Th. 10, p. 18] that, if $(a, m) = 1$, we can determine integers $u_0$ and $v_0$ such that

$$av_0 + m \cdot v_0 = 1$$

If we take $x_0 = -bu_0$, then

$$ax_0 + b = -abu_0 + b = b(-au_0 + 1) = b(mv_0) = (bv_0) \cdot m$$

which implies $ax_0 + b \equiv 0(\text{mod}\, m)$; so $x_0 = -bu_0$ is a solution of (21). If $x'$ is any other solution, then

$$ax' + b \equiv 0 \equiv ax_0 + b, \quad ax' \equiv ax_0(\text{mod}\, m)$$

Since $(a, m) = 1$, the common factor $a$ here can be cancelled, thus $x' = x0(\text{mod}\, m)$. Thereby I$^0$ is proved.

Assume that $g = (a, m) > 1$ and $g|b$. We have

$$a = ga_0, \quad b = gb_0, \quad m = gm_0 \text{ and } (a_0, m_0) = 1$$

$$g(a_0x + b_0) \equiv 0(\text{mod}\, gm_0)$$

which implies [See Th. 2,IV, p. 40]

(23) $$a_0x + b_0 \equiv 0(\text{mod}\, m_0), \quad a \leq x_1 \leq m_0$$

By I$^0$ the congruence (23) has a unique (mod $m_0$) solution $x_1$, thus

$$a_0x_1 + b_0 \equiv 0(\text{mod}\, m_0), \quad 0 \leq x_1 < m_0$$

*Claim:* $x_1, x_1 + m_0, x_1 + 2m_0, \ldots, x_1 + (g-1)m_0$
are all the distinct (mod $m$) solutions of the congruence (21)

It is clear that all these $g$ numbers satisfy the congruence (20): $ax + b \equiv 0(\text{mod}\, m)$. In fact

$$a(x_1 + tm_0) + b = ga_0(x_1 + tm_0) + gb_0$$

(24) $$= g(a_0x_1 + b_0) + tm_0 \cdot gm_0$$

Since $x_1$ is the solution of the congruence $a_0x + b_0 \equiv 0(\text{mod}\, m_0)$ we can write

$$a_0x_1 + b_0 = l_0m_0$$

Hence (24) becomes

$$a(x_1 + tm_0) + b = g(l_0 m_0) + a_0 g m_0 = (l_0 + a_0)m \equiv 0 (\text{mod}\, m)$$

i.e. for any $t$, $x_1 + tm_0$ is a solution of (21)

In order to find all the distinct mod $m$ solutions we determine the possible choices for $t$ such that

$$0 \leq x_1 + tm_0 \leq m - 1$$

We have

$$tm_0 \leq x_1 + tm_0 \leq gm_0 - 1 \quad t \leq g - \frac{1}{m_0}$$

Since $t$ is an integer this means that $t \leq g - 1$. Consequently

$$x_1 + tm_0, \quad t = 1, 1, \ldots, g - 1$$

give all the distinct mod m solutions of (21).

An important special case will be separately stated.

**Theorem 6.**

*If* $(a, m) = 1$, *there exists a unique* $a'$ *such that*

$$\boxed{a \cdot a' \equiv 1 (\text{mod}\, m)} \quad (1 \leq a' \leq m)$$

This is immediate since, under the condition $(a, m) = 1$, the congruence $ax \equiv 1 (\text{mod} \;\; m)$ has one and only solution $x$ in the interval $1 \leq x \leq m$.

We call $a'$ the mod $m$ inverse of $a$

**Example 8**

Given a modulus $M$ with $(10, M) = 1$, let $D$ denote the mod $M$-inverse of 10. Verify that

| $M$ | 11 | 23 | 17 | 19 |
|---|---|---|---|---|
| $D$ | 10 | 7 | 12 | 2 |

(i) If $D = 10$, $M = 11$, $10D = 100 \equiv 1 (\text{mod}\, M)$

(ii) If $D = 7$, $M = 23$, $10D = 70 \equiv 1 (\text{mod}\, M)$

(iii) If $D = 12$, $M = 17$, $10D = 120 \equiv 1 (\text{mod}\, M)$

(iv) If $D = 2$, $M = 19$, $10D = 20 \equiv 1 (\text{mod}\, M)$

Herein the modulus $M$ must satisfy the condition $(10, M) = 1$, i.e. $M$ is neither divisible by 2 nor by 5.

We consider now simultaneous linear congruences and prove the General remainder theorem.

**Theorem 7. [General Remainder Theorem]**

*If $m_1, \ldots, m_k$ are natural numbers coprime in pairs:*

$$(m_i, m_j) = 1 for i \neq j;$$

*and $m = m_1, m_2, \ldots m_k$, the sytem of congruences*

$$x \equiv a_i (\text{mod}\, m_i) \quad (i = 1, \ldots k)$$

*has a unique* mod $m$ *solution.*

*Proof:* Since $M_1 = m_1 m_2 \ldots m_k$ and $m_1$ is coprime with each of the $m_2, m_3, \ldots, m_k$ it follows that $M_1$ and $m_1$ are corpime. Similarly $M_i$ and $m_i$ are coprime for $i = 2, \ldots, k$

$$(M_i, m_i) = 1 \text{ for } i = 1, \ldots, k$$

By Th. 6 there exist $M_i'(i = 1, \ldots, k)$ such that

$$M_i M_i' \equiv 1 (\text{mod}\, m_i) \quad (i = 1, \ldots, k)$$

Let us take

(25) $$x_0 = a_1 M_1 M_1' + \ldots + a_k M_k M_k'$$

Since each $M_i$ for $i \neq 1$ contains $m_1$ as a factor

$$M_i \equiv 0 (\text{mod}\, m_1) \quad (i = 2, \ldots k)$$

whence follows

$$M_i M_i' \equiv 0 (\text{mod}\, m_1)$$

$$a_i M_i M_i' \equiv 0 (\text{mod}\, m_1) \quad (i = 2, \ldots k)$$

(26) $$a_2 M_2 M_2' + \ldots + a_k M_k M_k' \equiv 0 (\text{mod}\, m_1)$$

Also,

$$M_1 M_1' \equiv 0 (\text{mod}\, m_1)$$

So that

(27) $$a_1 M_1 M_1' \equiv a_1 (\text{mod}\, m_1)$$

Adding the two congruences (26) and (27) we get

$$(a_1 M_1 M_1') + (a_2 M_2 M_2' + \ldots + a_k M_k M_k') \equiv a_1 + 0 \equiv a_1 (\text{mod}\, m_1)$$

This means, (See (25)),

$$x_0 \equiv a_1 (\text{mod}\, M_1)$$

In exactly the same manner

$$x_0 \equiv a_i (\text{mod}\, M_i) \quad i = 2, \ldots, k$$

This completes the proof ■

**Example 9** [Bhaskara I] [522 AD]

Find that number which divided by 8 leaves 5 as remainder, divided by 9 leaves 4 as remainder and divided by 7 leaves 1 as remainder.

We have to solve the system

$$x \equiv 5 (\text{mod}\, 8), \quad x \equiv 4 (\text{mod}\, 9), \quad x \equiv 1 (\text{mod}\, 7)$$

As in Th. 7 we have: 8,9,7 are coprime in pairs.

$$\begin{array}{lll} m_1 = 8 & M_1 = 9.7 & 63M_1' \equiv 1(\text{mod}\, 8) \\ m_2 = 9 & M_2 = 7.8 & 56M_2' \equiv 1(\text{mod}\, 9) \\ m_3 = 7 & M_3 = 8.9 & 72M_3' \equiv 1(\text{mod}\, 7) \end{array}$$

These become

$$-M_1' \equiv 1(\text{mod}\, 8), \quad 2M_2' \equiv 1(\text{mod}\, 9), \quad 2M_3' \equiv 1(\text{mod}\, 7)$$

Whence we obtain $M_1' = 7$, $M_2' = 5$, $M_3' = 4$. Hence

$$\begin{aligned} x_0 &= 5M_1M_1' + 4M_2M_2' + 1M_3M_3' \\ &= 5.63.71 + 4.56.5 + 1.72.4 = 3613 \end{aligned}$$

Thus

$$x_0 \equiv 3613 (\text{mod}\, m), \quad m = m_1m_2m_3 = 8.9.7 = 504$$

Since $3613 \equiv 85(\text{mod}\, 504)$ the solution of the system is given by

$$x_0 \equiv 85 (\text{mod}\, 504)$$

**Example 10** [Bhaskara I] (522 AD)

Tell me at once, O mathematician, that number which leaves unity as remainder when divided by any of the numbers from 2 to 6 but is exactly divisible by 7.

This very problem was treated by Ibn-al-Haitam (1000 AD) and Leonardo Fibonacci (1202 AD)

We have to solve

$$\begin{aligned} &x \equiv 1(\text{mod}\, 2), \quad x \equiv 1(\text{mod}\, 3), \quad x \equiv 1(\text{mod}\, 4), \\ &x \equiv 1(\text{mod}\, 5), \quad x \equiv 1(\text{mod}\, 6) \text{ and } x \equiv 0(\text{mod}\, 7) \end{aligned} \tag{28}$$

Here the moduli 2,3,4,5,6,7 are not coprime in pairs.

However

$$x - 1 \equiv 0(\text{mod}\, 2, 3, 4, 5, 6)$$

means that $x - 1$ is a common multiple of 2,3,4,5,6.

Consequently $x - 1$ must be divisible by the l.c.m. of 2,3,4,5,6 i.e.

$$x - 1 \equiv 0 (\text{mod } 60)$$

Hence the system (28) is equivalent to

$$(29) \qquad \begin{cases} x \equiv 1 (\text{mod } 60 \\ x \equiv 0 (\text{mod } 7) \end{cases}$$

From the second congruence we see that $x = 7t$ where $t$ is determined from

$$(30) \qquad 7t \equiv 1 (\text{mod } 60)$$

For this purpose we find [ Compare Example 2, p. 23] the simple continued fraction for $\frac{60}{7}$:

| | | | | |
|---|---|---|---|---|
| 7 | 60 | | | |
| 8 | 56 | | | |
| | 4 | 7 | | |
| | 1 | 4 | | |
| | | 3 | 4 | |
| | | 1 | 3 | |
| | | | 1 | 3 |
| | | | 3 | 3 |
| | | | | * |

| $a_n$ | 8 | 1 | 1 | 3 |
|---|---|---|---|---|
| $p_n$ | 8 | 9 | 17 | 60 |
| $q_n$ | 1 | 1 | 2 | 7 |

$$N = 3 \qquad \begin{vmatrix} p_{N-1} & p_N \\ q_{N-1} & p_N \end{vmatrix} = \begin{vmatrix} 17 & 60 \\ 2 & 7 \end{vmatrix} = -1$$

$$7.17 - 2.60 = -1$$

$$7.17 \equiv -1 (\text{mod } 60)$$

This shows that the congruence (30) is satisfied by $t = -17$. Since $-17 \equiv 43 (\text{mod } 60)$ we find that the congruence (30) has the solution

$$t = 43$$

and therefore $x = 7t = 301$. Consequently the solution of the system (29) is given by

$$x \equiv 301 (\text{mod } 420)$$

The Th. 7 proved above and illustrated in Examples 9 and 10 is usually referred to as the Chinese Remainder Theorem . The following problems were considered in ancient Chinese publications that are dated 1295 A.D.

$1^0$ What is the number which leaves the remainders 1,2, and 3 when divided respectively, by 7,8 and 9?

$2^0$ What is the number which leaves the remainders 3,2, and 1, respectively, when divided by 11,12 and 13?

$3^0$ What is the number which leaves the remainders 3,2, and 1, respectively, when divided by 2,5 and 7?

$4^0$ (1593 AD)

*Three* people walking together, 'tis rare that one be *seventy*.
*Five* cherry blossom trees, *twentyone* branches bearing flowers,
*seven* disciples reunite for the *half-moon*.
Take away multiple of *105* and you shall know.

The problem was to solve:

$$x \equiv 2 (\bmod 3), \quad x \equiv 3 (\bmod 5), \quad x \equiv 2 (\bmod 7)$$

If $a$, $b$, $c$ were the remainders the rule of the song says:

Form the sum $x = 70a + 21b + 15c$.

In our specific case

$$a = 2, \quad b = 3, \quad c = 2 \quad 70a + 21b + 25c = 233$$

70 is a multiple of 5 and 7 such that $70 \equiv 1 (\text{mod } 3)$
21 is a multiple of 3 and 7 such that $21 \equiv 1 (\text{mod } 5)$
15 is a multiple of 3 and 5 such that $15 \equiv 1 (\text{mod } 7)$

Consequently

$$x = 70a + 21b + 25c \equiv \left| \begin{array}{l} a (\bmod 3) \\ b (\bmod 5) \\ c (\bmod 7) \end{array} \right.$$

$5^0$ (Ancient Chinese publiciations 1275 AD)
Seven with remainder 1, eight with remainder 2,
nine with remainder 3, What is the number?

However the General Problem of remainders was extensively studied by Aryabhata I (499 AD), Bhaskara I (522 AD)—the direct disciple and earliest commentator of Aryabhata I and by Brahmagupta (622 AD). Sripati (1039 AD), Bhaskara II (1150 AD) have contributed to this problem.

CHAPTER II

# TOPICS IN SRI BHARATHI KRISHNA THIRTHA'S VEDIC MATHEMATICS

## § 1. SOME QUESTIONS OF DIVISIBILITY

We shall show in the succeeding sections various applications of congruences: these will include simple direct proofs of several results of *Shri Bharathi Krishna Thirtha* as developed in his well-known book: **"VEDIC MATHEMATICS"** Motilal Banarsi Das (1978). In this section we consider his results presented in Chapter XXIX: Divisibility and simple osculators (See pp.273-284)

We recall that the congruence

$$ax \equiv b (\text{mod}\, m)$$

has a *unique* solution $x_0$, with $1 \leq x_0 \leq m-1$, provided the coefficient of $x$ is coprime with the modulus $m$, i.e. $(a, m) = 1$. In particular, taking $b = 1$, we see that

Given $a$, with $(a, m) = 1$, there exists one and only $a'$, with $1 \leq a' \leq m-1$ such that

$$aa' \equiv 1 (\text{mod}\, m) \tag{1}$$

In particular, when $m = p$ is a *prime*, for given a with $1 \leq a \leq p-1$ there exists a uniquely determined $a'$ with $1 \leq a' \leq p-1$ such that

$$aa' \equiv 1 (\text{mod}\, p) \tag{2}$$

We call $a'$ the *mod. p-inverse* of a

**Example 1.** $p = 11$ $\qquad aa' \equiv 1 (\text{mod}\, 11)$

| $a$ | 1 | 2 | 3 | 4 | 5 | 6 | 7 | 8 | 9 | 10 |
|---|---|---|---|---|---|---|---|---|---|---|
| $a'$ | 1 | 6 | 4 | 3 | 9 | 2 | 8 | 7 | 5 | 10 |

$a$ coincides with $a'$ in two cases: $a = 1$, and $a = 10$. The remaining $p - 3$ numbers can be paired:

$$(a, a') = (2, 6), (3, 4), (5, 9), (7, 8)$$

In (2) $a$ coincides with $a'$ if and only if

$$a^2 \equiv 1(\text{mod}\, p), (a-1)(a+1) \equiv 0(\text{mod}\, p)$$

Since $p$ is a prime this implies: either $a-1 \equiv 0$ or $a+1 \equiv 0$, either $a \equiv 1$, or $a \equiv -1 \equiv p-1(\text{mod}\, p)$. Since $1 \leq a \leq p-1$ this means:

$$a = 1 \quad \text{or} \quad a = p-1$$

The remaining $p-3$ numbers, (we assume that $p \neq 2$),

$$2, 3, \ldots, p-2$$

can be paired into $\dfrac{p-3}{2}$ pairs $(a, a')$. Since by (2),

$$aa' \equiv 1(\text{mod}\, p)$$

we obtain

$$2.3\ldots(p-2) \equiv 1.1.\ldots 1\left(\frac{p-3}{2} \text{ times}\right)$$

so that

$$1.(2.3\ldots p-2).(p-1) \equiv p-1 \equiv -1(\text{mod}\, p)$$

We have obtained:

**Example 2.** (Wilson's Theorem)

If $p$ is a prime, then $(p-1)! \equiv -1(\text{mod}\, p)$. In the proof above we assumed that the prime $p \neq 2$. But for $p = 2$, the statement is trivial.

Assume that the modulus m is coprime with 10, $(10, m) = 1$. We first observe that, when $m$ is coprime with 10, $m$ must end with one of the digits

$$1, \;\; 3, \;\; 7, \;\; 9$$

If a number $A$ ends with the digit $a_0$:

$$A = \langle a_n, a_{n-1}, \ldots, a_2, a_1, a_0 \rangle_{10}$$

i.e.

$$A = a_n.10^n + a_{n-1}.10^{n-1} + \cdots + a^0$$

we at once see that

(3) $$\boxed{A \equiv a_0(\text{mod}\, 10)}$$

($a_0$ is the *last* digit, i.e. the digit in the unit place) i.e.

every number is congruent (mod 10) to its last digit.

**Example**

$$12345 \equiv 5 (\bmod 10), \quad 12345 \equiv 45 (\bmod 10^2),$$
$$12345 \equiv 345 (\bmod 10^3)$$

**Theorem 1.** *Given $m$ such that $(10, m) = 1$, put*

$$m' = \begin{cases} \frac{9m+1}{10} & \text{for } m \equiv 1 (\bmod 10) \\ \frac{3m+1}{10} & \text{for } m \equiv 3 (\bmod 10) \\ \frac{7m+1}{10} & \text{for } m \equiv 7 (\bmod 10) \\ \frac{m+1}{10} & \text{for } m \equiv 9 (\bmod 10) \end{cases} \tag{4}$$

Then

$$\boxed{10m' \equiv 1 (\bmod m)} \tag{4}$$

The proof is immediate from (4)

**Example 1.** $m'$ = the mod $m$ inverse of 10 : $10m' \equiv 1 (\bmod m)$.

(5)

| $m$ | 19 | 29 | 13 | 17 | 61 | 47 |
|---|---|---|---|---|---|---|
| $m'$ | 2 | 3 | 4 | 26 | 55 | 33 |

While studying questions of divisibility by $m$ (Chapter XXIX) and the recurring decimal expansions of $\frac{1}{m}$ (Chapter XXVI) Shri Bharathi Krishna Thirthaji essentially exploits the use of the inverse (mod 10) of $m$. Whenever $m'$ is much smaller than $m$ the computations get simplified. However this need not always happen. [See (5)].

**Example 2.** To show that $A = 2774$ is divisible by $p = 19$. Let $D$ be the inverse of 10 mod 19:

$$10D \equiv 1 (\bmod 19), \quad \underline{D = 2}$$

(Add 1 to 19 and divide by 10, or increase the digit 1 in the 10-place to 2: We have (all congruences being mod 19)

$$A = 2770 + 4, \quad DA \equiv 277 + 4D = 277 + 8$$
$$= 285 = 280 + 5$$

$$D^2 A \equiv 28 + 5D = 28 + 10 = 38 \equiv 0$$

i.e.

$$D^2 A \equiv 0 (\bmod p)$$

Since $D = 2$ is coprime with $p = 19$, $D^2$ can be cancelled, $A \equiv 0$ $(\bmod p)$, i.e. $A$ is divisible by $p$

**Theorem 2. (Bharathi Krishna Thirtha)**

*Given the number* $A = \langle a_n, a_{n-1}, \ldots a_0 \rangle_{10}$ *and the modulus* $m$ *such that* $(10, m) = 1$ *form the sequence*

$$b_1, \ldots, b_n \quad (0 \le b_i < m)(i = 1, \ldots n)$$

*as follows*

(6)
$$\begin{aligned} Da_0 + a_1 &\equiv b_1 (\text{mod}\, m) \\ Db_1 + a_2 &\equiv b_2 (\text{mod}\, m) \\ &\vdots \\ Db_{n-1} + a_n &\equiv b_n (\text{mod}\, m) \end{aligned}$$

*Claim:* $A \equiv 0 (\text{mod}\, m)$ if and only if $b_n \equiv 0 (\text{mod}\, m)$ (i.e. $b_n = 0$, since $0 \le b_n < m$)

**Proof**: Multiplying by 10 the defining equations (6), and using the fact that $10D \equiv 1 (\text{mod}\, m)$ we have

$$\begin{aligned} a_0 + 10a_1 &\equiv 10\, b_1 (\text{mod}\, m) \\ b_1 + 10a_2 &\equiv 10\, b_2 (\text{mod}\, m) \\ \vdots \quad &\vdots \quad \vdots \\ b_{n-1} + 10a_n &\equiv 10\, b_n (\text{mod}\, m) \end{aligned}$$

Multiply the second equation by 10, the third by $10^2, \ldots$, and the last by $10^{n-1}$ and add; one gets, observing that the terms containing $b_1, b_2, \ldots, b_{n-1}$ cancel, that

$$a_0 + 10\, a_1, + 10^2 a_2 + \ldots + 10^n a_n \equiv 10^n b_n (\text{mod}\, m)$$

i.e., $A \equiv 10^n b_n (mod m)$
whence follows that $A \equiv 0 (\text{mod}\, m) \iff 10^n b_n \equiv 0 (\text{mod}\, m)$

Since $(10, m) = 1$ this means:

A is divisible by $m$ if and only if $b_n$ is divisible by $m$. But $0 \le b_n < m$. Hence $b_n = 0$

This completes the proof.

**Example 3.** [See Example (12). P.278 ≪ Vedic Mathematics ≫]

Show that the number $A = 7145501$ is divisible by 179

Put $p = 179$, then $10D \equiv 1 (\text{mod}\, p)$ for $\underline{D = 18}$

| $n$ | 0 | 1 | 2 | 3 | 4 | 5 | 6 |
|---|---|---|---|---|---|---|---|
| $a_n$ | 1 | 0 | 5 | 5 | 4 | 1 | 7 |
| $b_n$ | | 18 | 150 | 20 | 6 | 109 | $\underline{0}$ |

$$\begin{aligned}
Da_0 + a_1 &= 18 = b_1 \\
Db_1 + a_2 &= 324 + 5 = 329 \equiv 150 = b_2 \\
Db_2 + a_3 &= 2700 + 5 = 2705 \equiv 20 = b_3 \\
Db_3 + a_4 &= 360 + 4 = 364 \equiv 6 = b_4 \\
Db_4 + a_5 &= 108 + 1 = 109 = b_5 \\
Db_5 + a_6 &= 1962 + 7 = 1969 \equiv 0 = b_6
\end{aligned}$$

(all congruences being mod 179)

In Th.2, in the congruences (6), we may replace $D$ by any number congruent $(\text{mod}\, m)$ to it. In particular we can consider $\Delta$ defined by

(7) $$\boxed{D + \Delta = m}$$

so that

$$D \equiv -\Delta (\text{mod}\, m)$$

Thus in Th. 2 we can replace D by $-\Delta$:

(8) $$\begin{aligned}
-\Delta a_0 + a_1 &\equiv b_1 (\text{mod}\, m) \\
-\Delta b_1 + a_2 &\equiv b_2 (\text{mod}\, m) \qquad 0 \le b_i < m (i = 1, \ldots m) \\
&\vdots \\
-\Delta b_{n-1} + a_n &\equiv b_n (\text{mod}\, m)
\end{aligned}$$

If $D$ is large, but $\Delta$ is small we make computations using $\Delta$. For example, when $m = 41$, one has $D = 37$ and $\Delta = 4$

We have

(9) $$\boxed{10D \equiv 1 (\text{mod}\, m) \text{ and } 10\Delta \equiv -1 (\text{mod}\, m)}$$

*Remark*: In order to avoid, as far as possible, computations with large numbers, we choose the smaller of $D$ and $\Delta$, and we can use negative residues also.

(i) Positive residues $(\text{mod}\, m)$

$$\{0, 1, 2, \ldots, m-1\}$$

(ii) Absolutely least residues $(\bmod m)$

$$\left\{-\frac{m-1}{2},\ldots,-1,0,1,\ldots\frac{m-1}{2}\right\} \text{ for } m \text{ odd}$$

and for $m$ even one of the two series

$$\left\{-\frac{m}{2}+1,\ldots,-1,0,1,\ldots,\frac{m}{2}\right\}$$

or

$$\left\{-\frac{m}{2},\ldots,-1,0,1,\ldots,\frac{m}{2}-1\right\}$$

**Example 4.** [Bharati Krishna Thirtha, Vedic Mathematics p. 283]

Is $A = 165763$ divisible by $p = 41$?

Here we have $D = 37$ and $\Delta = 4$. We choose $\Delta$:

$$\Delta = 4, \quad 10\Delta \equiv -1(\bmod 41)$$

| $n$ | 0 | 1 | 2 | 3 | 4 | 5 |
|---|---|---|---|---|---|---|
| $a_n$ | 3 | 6 | 7 | 5 | 6 | 1 |
| $b_n$ | | −6 | 31 | 4 | −10 | <u>0</u> |

$$\begin{aligned}
-\Delta a_0 + a_1 &= -12 + 6 = -6 = b_1\\
-\Delta b_1 + a_2 &= 24 + 7 = 31 = b_2\\
-\Delta b_2 + a_3 &= -124 + 5 = -119 \equiv 4 = b_3\\
-\Delta b_3 + a_4 &= -16 + 6 = -10 = b_4\\
-\Delta b_4 + a_5 &= 40 + 1 = 41 \equiv 0 = b_5
\end{aligned}$$

(Since $b_2 = 31 \equiv -10(\bmod 41)$ we can use $-10$ also)

It follows that $A \equiv 0(\bmod m)$

When numbers are large we can choose, instead of 10, a higher power of 10, say $10^2$ or $10^3$, as the scale.

**Example 5.** [Sri Bh. Kr. Th. Vedic Mathematics P.288]

Prove that $A = 106656874269$ is divisible by $p = 499$

Herein $p + 1 = 500 = 5.10^2$. Let $\underline{M = 5}$, then

$$10^2.M \equiv 1(\bmod p)$$

Express $A$ in the scale of $10^2$:

$$A = 69 + 42.10^2 + 87.10^4 + 56.10^6 + 66.10^8 + 10.10^{10}$$

| $n$ | 0 | 1 | 2 | 3 | 4 | 5 |
|---|---|---|---|---|---|---|
| $a_n$ | 69 | 42 | 87 | 56 | 66 | 10 |
| $b_n$ | | −112 | 26 | 186 | −2 | 0 |

$$Ma_0 + a_1 = 345 + 42 = 387 \equiv -112 = b_1$$
$$Mb_1 + a_2 = -560 + 87 \equiv -61 + 87 \equiv 26 = b_2$$
$$Mb_2 + a_3 = 130 + 56 = 186 = b_3$$
$$Mb_3 + a_4 = 930 + 66 = 996 \equiv -2 = b_4$$
$$Mb_4 + a_5 = -10 + 10 = 0 = b_5$$

Consequently $A \equiv 0 (\mathrm{mod}\, p)$

## §2. RECURRING DECIMALS

*Definition 1.* We fix a positive integer $m \geq 2$ and call it the base number. We put 1 for $0 \leq a_i < m (i = 1, \ldots h)$,

(1) $$A_0 = \cdot a_1 a_2 \ldots a_h = \frac{a_1}{m} + \frac{a_2}{m^2} + \ldots + \frac{a_h}{m^h}$$

and

$$A = \cdot \overline{a_1 a_2 \ldots a_h} = \left[\frac{a_1}{m} + \frac{a_2}{m^2} + \cdots + \frac{a_h}{m^h}\right] + \left[\frac{a_1}{m^{h+1}} + \frac{a_2}{m^{h+2}} + \cdots + \frac{a_h}{m^h}\right] + \cdots$$

thus

(3) $$A = A_0 + \frac{A_0}{m^h} + \frac{A_0}{m^{2h}} + \cdots$$

Then $A$ is called a *recurring m-adic fraction* with respect to the base $m$ when we want to emphasize the base number we write

$$\left. \cdot \overline{a_1 a_2 \ldots a_h} \right|_{(m)}$$

Usually the base is taken to be $m = 10$.

The series in (3) is a geometric series of common ration $\frac{1}{m^h}$. Since $m \geq 2$ we have $\frac{1}{m^h} < 1$, and the series converges:

$$A = A_0 \frac{1}{1 - \frac{1}{m^h}} = A \frac{m^h}{m^h - 1}$$

From (1)

$$A_0.m^h = a_1.m^{h-1} + a_2.m^{h-2} + \ldots + a_{h-1}.m + a_h$$

Thus, when $m = 10$,

$$A_0.10^h = a_1.10^{h-1} + a_2.10^{h-2} + \ldots + a_{h-1}.10 + a_h$$

i.e.

$$A_0 10^h = \langle a_1, a_2, \ldots, a_{h-1}, ah\rangle_{(10)}$$

the number with the digits in the decimal scale.

Returning to the general base we write

(4) $$\boxed{\left.\cdot\overline{a_1 a_2 \ldots a_h}\right|_{(m)} = \frac{\langle a_1, a_2, \ldots, a_h\rangle_{(m)}}{m^h - 1}}$$

**Example 1.** Decimal expansion of $\frac{1}{7}$

| 1 | 4 | 2 | 8 | 5 | 7 | |
|---|---|---|---|---|---|---|
| 10 | 30 | 20 | 60 | 40 | 50 | 10 |
| 7 | 28 | 14 | 56 | 35 | 49 | |
| 3 | 2 | 6 | 5 | 5 | 1 | |

| 7 | 10 | .142857 |
|---|---|---|
| | 7 | |
| | 30 | |
| | 28 | |
| | 20 | |
| | 14 | |
| | 60 | |
| | 56 | |
| | 40 | |
| | 35 | |
| | 50 | |
| | 40 | |
| | 1 | |

Divide 10 by 7 to obtain the remainder 3;
Divide 30 by 7 to obtain the remainder 2;
and so on. At the very first step where we obtain the remainder 1 we are back to where we started from, i.e. the recurrence starts. We arrange the computations either in a horizontal or vertical table.

$$\frac{1}{7} = \cdot\overline{142857}$$

Since $\frac{1}{2} = \cdot 5$ and $\frac{1}{5} = \cdot 2$, we assume that $P$ has neither 2 nor 5 as a factor, i.e. we assume that

(5) $$\boxed{(10, p) = 1}$$

Our purpose is to obtain the recurring decimal expansion of $\frac{1}{P}$ under the assumption (5), without going through the tedious process of various divisions.

The central result in "VEDIC MATHEMATICS" Chapter XXVI. Recurring decimals, pp 196-239 is the following proposition:

**Theorem 1.** [Shri Bharathi Krishna Thirthaji]

*Given P with $(10, p) = 1$, the digits $a_k$ in the recurring decimal expansion of $\frac{1}{P}$:*

(6) $$\frac{1}{P} = \overline{\cdot a_1 a_2 \ldots a_N}$$

*are obtained as follows:* $1^0$ *determine the least positive residues* $r_k$ *of* $10^k \pmod{p}$:

$$\boxed{10^k \equiv r_k \pmod{p},} \qquad (0 \le r_k < p)$$

$2^0$ *Put*

$$\delta = \begin{vmatrix} 1 & \text{for } P \equiv 9 \pmod{10} \\ 3 & \text{for } P \equiv 3 \pmod{10} \\ 7 & \text{for } P \equiv 7 \pmod{10} \\ 9 & \text{for } P \equiv 1 \pmod{10} \end{vmatrix}$$

thus in all cases $P \cdot \delta \equiv 9 \pmod{10}$, or $P\delta \equiv -1 \pmod{10}$

$3^0$ *Then*

$$\boxed{\delta r_k \equiv a_k \pmod{10},} \qquad (0 \le a_k \le 9)$$

*i.e.* $a_k$ *are the last digits of* $\delta r_k$.

*Remarks*: Recall that $x \equiv a \pmod{10}$, $0 \le a \le 9$, means that $a$ is the last (i.e. unit) digit of 10. Really speaking, in (7) we are more interested in the last digits of $r_k$ than in the least positive residues $r_k$ themselves.

**Proof**: As we have seen in Example 1 the coefficients $a_k$ and the remainders $r_k$ are connected by the chain of equations.

(8)
$$\begin{aligned} 10 &= Pa_1 + r_1 \\ 10r_1 &= Pa_2 + r_2 \\ &\vdots \\ 10r_{k-1} &= Pa_k + r_k \\ &\vdots \\ 10r_{N-1} &= Pa_N + r_N \end{aligned} \qquad \begin{aligned} &0 \le r_k < P \quad (k = 1, 2, \ldots N^{-1}) \\ &r_N = 1 \end{aligned}$$

It is at once seen, by induction on $k$, that

$$(9) \qquad 10^k \equiv r_k (\text{mod}\, P)$$

In fact, from the first equation in (8), $10 \equiv r_1 (\text{mod}\, p)$ which proves (9) for $k = 1$. Assume (9) for $k - 1$:

$$(10) \qquad 10^{k-1} \equiv r_{k-1} (\text{mod}\, P)$$

From (8) we have $10r_{k-1} = Pa_k + r_k$ so that

$$r_k \equiv 10r_{k-1} (\text{mod}\, P)$$

so that, owing to the inductive assumption (10)

$$r_k \equiv 10.10^{k-1} \equiv 10^k (\text{mod}\, p)$$

which completes the induction and establishes (7)

Again, from (8), we have $Pa_k + r_k \equiv 0 (\text{mod}\, 10)$ so that

$$r_k \equiv -Pa_k (\text{mod}\, 10)$$

Since $\delta r_k$ is so chosen in $2^0$ that $P\delta \equiv -1 (\text{mod}\, 10)$ we get

$$\delta \equiv -P\delta.a_k \equiv a_k (\text{mod}\, 10)$$

which yields $3^0$ and completes the proof of the theorem. ■

*Remark 1.* We must observe that there always exists a positive integer $N$ such that

$$10^N \equiv 1 (\text{mod}\, P), \quad (10, P) = 1$$

This is secured by Euler-Fermat theorem since

$$10^{\varphi(P)} \equiv 1 (\text{mod}\, P)$$

We can say further:

(11) The least positive integer $N$ with the property $10^N \equiv 1 (\text{mod}\, P)$ must always be a divisor of $\varphi(P)$

In fact, dividing $\varphi(P)$ by $N$:

$$\varphi(P) = q.N + r, \qquad 0 \leq r < N$$

Then

$$1 \equiv 10^{\varphi(P)} = 10^{qN+r} = (10^N)^q.10^r \equiv 1.10^r$$

i.e., $10^r \equiv 1 (\text{mod}\, P)$. This implies that $r = 0$, otherwise $r$ would be a positive integer with the property (11), contradicting the minimality of $N$. In this manner

$$\varphi(P) = q.N, N|\varphi(P)$$

We call $N$ the *period (mod P) of 10*. The number of digits in the recurring decimal expansion of $\frac{1}{P}$ is precisely equal to $N$.

*Remark 2*. In computations it is often convenient to use, instead of $10^k (k = 1, 2, \ldots)$, the sequence

$$\{D^k\}(k = 1, 2, \ldots) \text{ where } 10D \equiv 1 (\text{mod}\, P)$$

i.e. $D$ is the inverse of $10 (\text{mod}\, P)$. If $N$ is the period $(\text{mod}\, P)$ of 10 we have

$$10^N \equiv 1 (\text{mod}\, P)$$

Since we have $10^k \equiv r_k (\text{mod}\, P)$, $(k = 1, 2, \ldots N)$, we get $10^N \equiv 1 \equiv (10D)^k \equiv 10^k.D^k$, i.e. $10^k.10^{N-k} \equiv 10^k.D^k (\text{mod}\, P)$

Since $(10, P) = 1$, the common factor $10^k$ here can be cancelled to yield

(12) $$10^{N-k} \equiv D^k (\text{mod}\, P)$$

which implies that $10^{N-k}$ and $D^k$ must have the same least positive residue $(\text{mod}\, P)$. Denote by $\Delta_k$ the least positive residue $(\text{mod}\, p)$ of $D^k$:

(13) $$D^k \equiv \Delta_k (\text{mod}\, P), \quad 0 < \Delta_k < P$$

From (12) and (13) we have

| $a_1$ | $a_2$ | $\cdots$ | $a_k$ | $a_n$ |
|---|---|---|---|---|
| $\delta\Delta_{N-1}$ | $\delta\Delta_{N-2}$ | $\cdots$ | $\delta\Delta_{N-k}$ | $\delta\Delta_0$ |
| $\delta r_1$ | $\delta r_2$ | $\cdots$ | $\delta r_k$ | $\delta r_N$ |

$$r_N = \Delta_o = 1, \quad 10^N \equiv 1 (\text{mod}\, P)$$

$$P\delta \equiv -1 (\text{mod}\, 10)$$

$$\delta = \begin{cases} 1 & \text{for } P \equiv 9 (\text{mod } 10) \\ 3 & \text{for } P \equiv 3 (\text{mod } 10) \\ 7 & \text{for } P \equiv 7 (\text{mod } 10) \\ 9 & \text{for } P \equiv 1 (\text{mod } 10) \end{cases}$$

Let us call $\delta$ the *multiplier* for $P$. We have

$$\delta = 1, 3, 7, 9$$

according as the last digit of P is $9, 3, 7, 1$

(14) $$10^k \equiv r_k (\text{mod } p), 0 < r_k < P, (k = 1, 2, \ldots N)$$

$$\boxed{r_k = \Delta_{N-k}} \quad (k = 1, \ldots, N)$$

*Remark 3*. The following observation will be useful when we compute the least positive residues of $D^k$. If $x \equiv \langle x_1, x_0 \rangle_{10} = 10x_1 + x_0 (\text{mod } P)$, then

$$Dx \equiv x_1 + Dx_0$$

In this manner, if for example, $P = 19$, (then $D = 2$), and $x = 14$,

$$Dx \equiv 1 + 4D = 9$$

In fact, since $10D \equiv 1 (\text{mod } P)$,

$$Dx = D(10x_1 + x_2) = 10D.x_1 + Dx_2 \equiv x_1 + Dx_2 (\text{mod } P)$$

**Example 2.** $P = 19$

We have $D = 2, \delta = 1, N = 18$

| $k$ | 1 | 2 | 3 | 4 | 5 | 6 | 7 | 8 | 9 | 10 | 11 | 12 | 13 | 14 | 15 | 16 | 17 | 18 |
|---|---|---|---|---|---|---|---|---|---|---|---|---|---|---|---|---|---|---|
| $\Delta_{N-k}$ | 10 | 5 | 12 | 6 | 3 | 11 | 15 | 17 | 18 | 9 | 14 | 7 | 13 | 16 | 8 | 4 | 2 | 1 |
| $a_k$ | 0 | 5 | 2 | 6 | 3 | 1 | 5 | 7 | 8 | 9 | 4 | 7 | 3 | 6 | 8 | 4 | 2 | 1 |

(15) $$\boxed{\frac{1}{19} = \cdot\overset{*}{0}52631578/94736842\overset{*}{1}}$$

(The asterisk is put to indicate recurrence.)

The computations are made as explained in Remark 3. For example $\Delta_4 = 16$, i.e.

$$D^4 = 16, \text{ so that } D^5 \equiv 1 + 6D = 13, \text{ thus } \Delta_5 = 13$$

*Remark 4.* We observe in the above example that

$$D^9 = 2^9 \equiv 18 \equiv -1 (\text{mod } 19)$$

so that, for $k = 1, 2, \ldots, 9$

(16) $$D^9 . D^k \equiv -D^k (\text{mod } 19)$$

Since $D^9 . D^k = D^{9+k} \equiv \Delta_{9+k}$ and $-D^k \equiv -\Delta_k \equiv 19 - \Delta_k$ we get

(17) $$\Delta_{9+k} = 19 - \Delta_k,$$

since the least positive residues on both sides of the congruence (16) must coincide.

If we now take congruence mod 10 of both sides in (17), (i.e. if we compare the last digits of both the numbers in (17) we get

(18) $$\boxed{a_{9+k} = 9 - a_k}$$

i.e.,

(*) | The decimal expansion (15) has the property of symmetry (18), namely: the last nine digits are obtained as complementaries in (9) of the first nine digits.

*Remark 5.* The decimal expansion of $\frac{1}{P}$ has the property of symmetry (*) in (18). Whenever the period $N$ is an *even* integer. *(P Prime).*

In fact, if $N = 2N'$ is even, we have

$$10^{2N'} = 10^N \equiv 1 (\text{mod } P)$$

$$(10^{N'} - 1)(10^{N'} + 1) \equiv 0 (\text{mod } P)$$

Since $P$ is assumed to be a prime, this implies:

$$\text{either } 10^{N'} - 1 \equiv 0 (\text{mod } P) or 10^{N'} + 1 \equiv 0 (\text{mod } P)$$

Herein the first alternative cannot happen, for otherwise

$$10N' \equiv 1 (\text{modP})$$

contradicting that $N$ is the smallest positive integer with the property $10^N \equiv 1(\text{modP})$. Accordingly

$$10^{N'} \equiv -1 (\text{modP})$$

thus,

$$\left. \begin{array}{c} P \text{ Prime} \\ \text{period } N \text{ is even} \end{array} \right| \Rightarrow 10^{N/2} \equiv -1 (\text{mod } P)$$

We now give an example where the period $N$ is an *odd* integer, (and therefore the symmetry property (18) does not hold.)

**Example 3.** $\underline{P = 31}$, Since $P \equiv 1 (\text{mod}\, 10)$ we must have $\underline{\delta = 9}$ Since $9P + 1 = 279 + 1 = 280$, we have

$$\underline{10D \equiv 1 (\text{mod}\, P)} \text{ with } \quad \underline{D = 28}$$

In order to find the least positive residues $\Delta_k of D^k (\text{mod}\, P)$ we need the residues of multiples of $D$:

$$k.D(k = 1, 2, \ldots, 9)$$

We have, all congruences being (mod 31)

| | | |
|---|---|---|
| $D = 28$ | $5D = D + 4D \equiv 28 + 19$ | $8D \equiv 2.4D$ |
| $2D = 56 \equiv 25$ | $\equiv 47 \equiv 16$ | $\equiv 38 \equiv 7$ |
| $3D \equiv 28 + 25$ | $6D \equiv 2.3D$ | $9D \equiv D + 8D$ |
| $= 53 \equiv 22$ | $\equiv 44 \equiv 13$ | |
| $4D \equiv D + 3D$ | $7D \equiv 2D + 5D \equiv 41$ | $\equiv 35 \equiv 4$ |
| $\equiv 28 + 22$ | | |
| $\equiv 50 \equiv 19$ | $\equiv 10$ | |

In this manner we have obtained the table of residues (mod 31) of k.D for $1 \le k \le 9$

(19)

| $D$ | $2D$ | $3D$ | $4D$ | $5D$ | $6D$ | $7D$ | $8D$ | $9D$ |
|---|---|---|---|---|---|---|---|---|
| 28 | 25 | 22 | 19 | 16 | 13 | 10 | 7 | 4 |

Recall that the computations of the least positive residues of $\Delta_k$ of $D^k (\text{mod}\, P)$ depend on the basic rule (See Remark 3, p. 70) If $x = 10x_1 + x_0$, $(0 \le x_0 \le 9)$, then $D_x \equiv x_1 + Dx_0$. Herein, since $0 \le x_0 \le 9$, we would require the least positive residues $(\text{mod}\, P)$ of $Dx_0$ obtained in (19).

We shall exhibit the first few computations.

Since $10D \equiv -1 (\text{mod}\, P)$, we have, successively, (using (19)):

$D^2 = D.D = 28.D \equiv 2 + 8D \equiv 2 + 7 = 9$

$D^3 = D.D^2 \equiv 9D \equiv 4; \quad D^4 \equiv 4D \equiv 19; \quad D^5 \equiv 19D \equiv 1 + 9D \equiv 5$

And so on.

| $k$ | 1 | 2 | 3 | 4 | 5 | 6 | 7 | 8 | 9 | 10 | 11 | 12 | 13 | 14 | 15 |
|---|---|---|---|---|---|---|---|---|---|---|---|---|---|---|---|
| $\Delta_{N-k}$ | 10 | 7 | 8 | 18 | 25 | 2 | 20 | 14 | 16 | 5 | 19 | 4 | 9 | 28 | 1 |
| $a_N$ | 0 | 3 | 2 | 2 | 5 | 8 | 0 | 6 | 4 | 5 | 1 | 6 | 1 | 2 | 9 |

$$a_k \equiv \delta\Delta_{N-k} \equiv 9\Delta_{N-k} (\text{mod}\, 10)$$

The least positive integer $N$ with the property $D^N \equiv 1(\text{mod}\, P)$ is obtained for $N = 15$ (which is *odd*). The symmetry property (18) does not hold. Also observe [See Remark 1, p. 68] that $N$ is a divisor of $P - 1 = 30$

$$\boxed{\frac{1}{31} \cdot \overset{*}{0}3225806451612\overset{*}{9}}$$

If we assume that $P$ is a *prime*, the numbner $N$ of digits in the decimal expansion of $\frac{1}{P}$ is the least positive integer $N$ such that $10^N \equiv 1(\text{mod}\, P)$, and $N$ is a divisor of $P - 1$. The maximum possible value that it can have is therefore $P - 1$. By Fermat's theorem

$$10^{P-1} \equiv 1(\text{mod}\, P) \text{ provided } P \dagger 10$$

If $P$ is such that for no lower power this happens, i.e.

$$10^r \not\equiv 1(\text{mod}\, P) \text{ for } r = 1, 2, \ldots, P - 2$$

then we say that 10 is a *primitive* root of $P$.

We state the following result [See L.K.HUA:Introduction to Number Theory, Springer verlag (1982), PP 52-56]

$1^0$ Among the primes $P$ with $P < 500$ the following primes have 10 as a primitive root:

$P =$ 7, 17, 19, 23, 29, 47, 59, 61, 97, 109, 113, 131, 149, 167, 179, 181, 193, 223, 229, 233, 257, 263, 269, 313, 337, 367, 379, 383, 389, 419, 433, 461, 487, 491, 499.

$2^0$ Among the primes $P$ with $4000 < P < 5000$ the following primes have 10 as a primitive root:

$P =$ 4007, 4019, 4051, 4057, 4073, 4091, 4139, 4153, 4177, 4211, 4217, 4219, 4229, 4259, 4261, 4327, 4339, 4349, 4421, 4423, 4447, 4451, 4457, 4463, 4567, 4583, 4651, 4673, 4691, 4703, 4783, 4793, 4817, 4931, 4937, 4943, 4967

Hence, for example, for the decimal expansion of $\frac{1}{4967}$, the recurrence starts after 4966 steps. The expansion has 4966 digits.

**Example 4.** $\underline{P = 89}$ We have $D = 9$ and $\delta = 1$

We write down the least positive residues of

$$D^{N-k} \quad (k = 1, 2, \ldots N)$$

(20)

| | | |
|---|---|---|
| $D \equiv 9$ | $D^{11} \equiv 34$ | $D^{21} \equiv 79$ |
| $D^2 \equiv 81$ | $D^{12} \equiv 39$ | $D^{22} \equiv 88$ |
| $D^3 \equiv 17$ | $D^{13} \equiv 84$ | |
| $D^4 \equiv 64$ | $D^{14} \equiv 44$ | |
| $D^5 \equiv 42$ | $D^{15} \equiv 40$ | |
| $D^6 \equiv 22$ | $D^{16} \equiv 4$ | |
| $D^7 \equiv 20$ | $D^{17} \equiv 36$ | |
| $D^8 \equiv 2$ | $D^{18} \equiv 57$ | |
| $D^9 \equiv 18$ | $D^{19} \equiv 68$ | |
| $D^{10} \equiv 73$ | $D^{20} \equiv 78$ | |

Since $D^{22} \equiv -1 (\bmod P)$ the property of symmetry

Now

$$a_k = \delta \Delta_{N-k}$$

and

$$\Delta_k + \Delta_{22+k} = 89$$

$$(k = 1, 2, \ldots, 22)$$

$$a_k + a_{22+k} = 9$$

The right sides of the congruences (20) are the least positive residues $\Delta_k (1 \leq k \leq 22)$. Since $\delta = 1$ we have

$$a_k \equiv \Delta_{44-k} (\bmod 10)$$

Consequently we obtain

$$\boxed{\frac{1}{89} = \begin{array}{l} \cdot * 0112\ 3595\ 5056\ 1797\ 7528\ 08\,| \\ |9887\ \ 6404\ 4943\ 8202\ 2471\ 91* \end{array}}$$

Since for $k = 0$, $D^k \equiv \Delta_k$ always implies $\Delta_0 = 1$, the last digits of the residues

$$\{\Delta_k\} \qquad k = 0, 1, 2, \ldots, 21$$

give the coefficients $a_k (k = 23, \ldots, 44)$ i.e. the second half of the expansion (21); the property of symmetry then completes the first half.

**Example 5.** $\underline{P = 49}$

Now $P$ is not a prime. We have $D = 5$, and $\delta = 1$, also

$$\varphi(P) = \varphi(7^2) = 7^2 (1 - \frac{1}{7}) = 42$$

It can be verified that the period $N = 42$. Since

$$D^{21} \equiv -1 (\bmod P)$$

The least positive residues $\Delta_k$ are given by

$$D^k \equiv \Delta_k (\text{modP}) \quad (k = 0, 1, 2, \ldots 21)$$

$$\Delta_k = \left| \begin{array}{l} 1, 5, 25, 27, 37, 38, 43, 19, 46, 34, 23, \\ 17, 36, 33, 18, 41, 9, 45, 29, 47, 39|\underline{48} \end{array} \right.$$

$$D^{21} \equiv 48 \equiv -1(\text{mod}\, P), \quad \Delta_{21} = 48$$

Each term is the sequence $\{\Delta_k\}$ is obtained from the preceding according to the rule

$$10x_1 + x_0 \longrightarrow Dx_0 + x_1$$

(whenever $Dx_0 + x_1$ exceeds $P$ we reduce it mod $P$). Accordingly

$$\frac{1}{49} = \cdot\overset{*}{0}204 \quad 0816 \quad 3265 \quad 3061 \quad 2244 \quad 8$$
$$9795 \quad 9183 \quad 6734 \quad 6938 \quad 7755 \quad \overset{*}{1}$$

**Example 6.** $\underline{P = 73}$

We have $D = 22, \quad \delta = 3$

All congruences being mod $P$ we first have

| $D$ | $2D$ | $3D$ | $4D$ | $5D$ | $6D$ | $7D$ | $8D$ | $9D$ |
|---|---|---|---|---|---|---|---|---|
| 22 | 44 | 66 | 15 | 37 | 59 | 8 | 30 | 52 |

Using this we have $D^0 = 1, \quad D = 22$, and

$$D^2 \equiv 2 + 2D \equiv 46,\ D^3 \equiv 4 + 6D \equiv 63,\ D^4 \equiv 6 + 3D \equiv 72$$

Thus

$$D^4 \equiv -1(\text{mod}\, P)$$
$$D^8 \equiv 1(\text{mod}\, P)$$

The expansion has $N = 8$ digits ($N$ is a division of $P - 1 = 72$). The coefficients of expansion are given by

$$a_{8-k} \equiv \delta D^k(\text{mod}\, 10) \qquad (k = 0, 1 \ldots, 7)$$

Hence

$$\frac{1}{73} = \cdot\overset{*}{0}136986\overset{*}{3}$$

## §3. SQUARE

Given a number in decimal notation

$$X = x_0 + x_1.10 + x_2.10^2 + \ldots + x_n.10^n$$
$$0 \le x_i \le 9 (i = 0, 1, \ldots n), x_n \ne 0$$

We want to relate the number $X^2$ to the digits $x_0, x_1, \ldots x_n$ of $X$. For this purpose it is convenient to make use of the following notation.

*Definition 1 (≪ Duplex ≫)*

Given the numbers $x_0, x_1, x_2, \ldots$, we shall put $(x_0) = x_0^2, (x_0, x_1) = 2x_0x_1, (x_0, x_1, x_2) = x_1^2 + 2x_0x_2,$

$$(x_0, x_1, x_2, x_3) = 2x_0x_3 + 2x_1x_2$$
$$(x_0, x_1, x_2, x_3, x_4) = 2x_0x_4 + 2x_1x_3 + x_2^2$$

In general

(1) $$(x_0, x_1, \ldots, x_k, \ldots, x_{2k}) = x_k^2 + 2x_0x_{2k} + 2x_1x_{2k-1} + \ldots + 2x_{k-1}x_{k+1}$$

and

(2) $$(x_0, \ldots, x_k, x_{k+1}, \ldots, x_{2k+1}) = 2x_0x_{2k+1} + 2x_1x_{2k+1} + \ldots + 2x_kx_{k+1}$$

Thus in $(x_0, x_1, \ldots x_n)$ the number of terms is $= n+1$ which is odd when $n$ is even, and is even when $n$ is odd. Hence when $n$ is even we have only one middle term, namely $x_k (k = \frac{n}{2})$; and when $n$ is odd there will be two middle terms, namely $x_k$ and $x_{k+1} \left( n = 2k+1, \text{ thus } k = \frac{n-1}{2} \right)$

We call $(x_0, x_1, \ldots, x_n)$ the *duplex* of the system of numbers $x_0, x_1, \ldots, x_n$ *in that order.*

In terms of this notation we can write the square of $X$ as follows:

$$\begin{aligned} X^2 &= (x_0 + x_1.10 + x_2.10^2 + \ldots + x_n.10^n)^2 \\ &= x_0^2 + 2x_0x_1.10 + (2x_0x_2 + x_1^2)10^2 + 2(x_0x_3 + x_1x_2)10^3 + \ldots \\ &\quad + \ldots + 2x_{n-1}x_n10^{2n-1} + x_n^2.10^{2n} \end{aligned}$$

The coefficients here are the dupliexes introduced in (1) and (2). We put

(3) $$\left| \begin{array}{l} \xi_0 = (x_0) \\ \xi_1 = (x_0, x_1) \\ \vdots \\ \xi_n = (x_0, x_1, \ldots, x_n) \end{array} \right| \begin{array}{l} \xi_{n+1} = (x_1, \ldots, x_n) \\ \xi_{n+2} = (x_2, \ldots, x_n) \\ \vdots \\ \xi_{2n} = (x_n) \end{array}$$

thus

(4) $$\boxed{X^2 = \xi_0 + \xi_1.10 + \ldots + \xi_k 10^k + \ldots + \xi_{2n} x^{2n}}$$

where

$$X = x_0 + x_1.10 + \ldots + x_n.10^n$$
$$0 \leq x_k \leq 9(k = 0, 1, 2 \ldots n)$$
$$x_n \neq 10$$

It must be observed that the coefficients $\xi_0 \ldots, \xi_{2n}$ in (4) are not, generally speaking, the digits of $X^2$ since they will not lie always between 0 and 9.

The problem is now to find the digits of $X^2$.

If $1 \leq X < 10$, the number $X$ has one digit; if $10 \leq X \leq 10^2$ it has two digits, and

$$X \text{ has } n \text{ digits } \Leftrightarrow 10^{n-1} \leq X < 10^n$$

Then $10^{2n-2} \leq X^2 < 10^{2n}$ so that:

$$\text{either } 10^{2n-2} \leq X^2 < 10^{2n-1} \text{ or } 10^{2n-1} \leq X^2 < 10^{2n}$$

Hence $X^2$ will have either $2n$ or $2n - 1$ digits.

The number $X = x_0 + x_1.10 + x_2.10^2 + \ldots + x_n.10^n$, $(x_n \neq 0)$ has $n + 1$ digits, and

$$X \leq 9 + 9.10 + 9.10^2 + \ldots + 9.10^n$$
$$= 9(1 + 10 + 10^2 + \ldots 10^n)$$
$$= 9\frac{10(n+1) - 1}{10 - 1} < 10^{n+1}$$

Since $x_k$ are all $\geq 0$, $X \geq x_n.10^n \geq 10^n$ (since $x_n \geq 1$), thus

$$10^n \leq X < 10^{n+1}$$

We shall mention some simple properties of the duplexes for later use. Since the digits $x_0, \ldots, x_n$ satisfy the inequalities

$$0 \leq x_k \leq 9 \quad (k = 0, 1, \ldots n)$$

we have

$$\xi_0 = (x_0) = x_0^2 \leq 9^2$$
$$\xi_1 = (x_0 x_1) = 2x_0 x_1 \leq 2.9^2$$
$$(5) \qquad \xi_2 = (x_0 x_1 x_2) = 2x_0 x_2 + x_1^2 \leq 2.9^2 + 9^2 = 3.9.^2$$
$$\vdots$$
$$\xi_n = (x_0 x_1 \ldots x_n) \leq (n + 1).9^2$$

and

$$\xi_{n+1} = (x_1 \ldots x_n) \leq n.9^2$$

$$\xi_{n+2} = (x_2 \ldots x_n) \leq (n-1).9^2$$

(6) $$\vdots$$

$$\xi_{2n} = (x_n) \leq 9^2$$

We shall now prove the following theorem which gives the digits of $X^2$ by means of the duplexes of $X$.

*Preliminary Remarks:*

Given a positive integer

$$T = t_0 + t_1.10 + t_2.10^2 + \ldots + t_m.10^m \qquad (0 \leq t_k \leq 9)$$

we can put it in the form

(7) $$\boxed{T = t_0 + 10T',} \qquad T' = t_1 + t_2.10 + \ldots + t_m 10^{m-1}$$

Here $t_0$ is the "last" (i.e. the unit) digit of $T$, and $T'$ is what results from $T$ when the last digit to is struck off. For example:

(i) $T = 54321$, $T' = 5432$, $t_0 = 1$

(ii) $T = 510$, $T' = 51$, $t_0 = 0$

(iii) $T = 7$, $T' = 0$, $t_0 = 7$

We can also describe $t_0$ as the remainder and $T'$ the quotient that are obtained when $T$ is divided by 10:

(8) Divide $T$ by 10: quotient $= T'$

remainder $= t_0$

Accordingly, when it is a question of division by 10, the last digit is the remainder, and what remains after it is struck off is the quotient. Since

$$T = 10.T' + t_0, \qquad 0 \leq t_0 \leq 9$$

$$\frac{T}{10} = T' + \frac{t_0}{10}, \qquad 0 \leq \frac{t_0}{10} < 1$$

We can describe the quotient $T'$ as the integral part of $\frac{T}{10}$:

(9) $$\boxed{T' = \left[\frac{T}{10}\right]}$$

The decimal notation exhibits this directly: e.g

$$T = 54321, \quad \frac{T}{10} = 5432.1, \quad \left[\frac{T}{10}\right] = 5432$$

**Theorem 1. [Shri Bharathi Krishna Thirthaji]**

*Let $\xi_0, \ldots, \xi_n$ be the duplexes for the number*

$$X = x_0 + x_1.10 + \ldots + x_n.10^n$$

*Define, successively, $\xi_0', \xi_1', \ldots, \xi_{2n-1}', \xi_{2n}'$ as follows:*

(10) $$\xi_0' = \left[\frac{\xi_0}{10}\right], \text{ and for } k = 1, \ldots, 2n, \quad \xi_k' = \left[\frac{\xi_{k-1}' + \xi_k}{10}\right]$$

*(wherein $[\cdots]$ denotes, as usual, the integral part)*

*Put*

$$a_k = \left| \begin{array}{l} \text{the last digit of } \xi_0 \quad \text{for } k = 0 \\ \text{the last digit of } \xi_{k-1}' + \xi_k \quad \text{for } k = 1, \ldots, 2n \\ \xi_{2n}' \quad \text{for } k = 2n+1 \end{array} \right.$$

*Claim:* $a_0, a_1, \ldots, a_{2n}$ are the digits of $X^2$:

$$\boxed{X^2 = a_0 + a_1.10 + a_2.10^2 + \ldots + a_{2n}.10^{2n}}$$

i.e.

(12) $$0 \le a_k \le 9 \quad (k = 0, 1, \ldots 2n)$$

Since in (11), by very definition, $a_0, a_1, \ldots a_{2n-1}$ are the last digits of certain integers we have, at the outset

$$0 \le a_0, a_1, \ldots, a_{2n-1} \le 9$$

Since in (11) $a_{2n}$ was defined as $\xi_{2n}'$ the essential point is to show that $0 \le \xi_{2n}' \le 9$.

**Proof of Theorem 1**

We have from (5) and (6)

(13) $$\left.\begin{array}{l} \xi_0 \le 9^2 \\ \xi_1 \le 2.9^2 \\ \vdots \\ \xi_{n-1} \le n.9^2 \\ \xi_n \le (n+1).9^2 \end{array}\right| \begin{array}{l} \xi_{n+1} \le n.9^2 \\ \xi_{n+2} \le (n-1).9^2 \\ \vdots \\ \\ \xi_{2n} \le 9^2 \end{array}$$

By the definition of $\xi_0', \ldots, \xi_{2n}'$ in (10) we see that, using (13),

$$\xi_0' = \left[\frac{\xi_0}{10}\right] \le \left[\frac{9^2}{10}\right] = \left[\frac{81}{10}\right] = 8;$$

$$\xi_1' = \left[\frac{\xi_0+\xi_1}{10}\right] \le \left[\frac{8+2.9^2}{10}\right] = \left[\frac{170}{10}\right] = 17;$$

and in general, as can be verified by induction on $k$,

(14) $$\xi_k' \le 9k+8 \quad (k=0,1,\ldots,n)$$

(15) $$\xi_{n+k}' \le 9(n-k+1) \quad (k=1,2,\ldots,n)$$

In fact, we have already seen above, that $\xi_0' \le 8$ and $\xi_1' \le 17$, i.e. (14) holds for $k=0,1$. Assume (14) for $k-1$;

$$\xi_{k-1}' \le 9(k-1)+8 = 9k-1$$

then

$$\xi_k' = \left[\frac{\xi_{k-1}'+\xi_k}{10}\right] \le \left[\frac{(9k-1)+(k+1)9^2}{10}\right],$$

since, by (13), $\xi_k \le (k+1)9^2$

$$= \left[\frac{90k+80}{10}\right] = 9k+8$$

which completes the induction on $k$ and upholds (14).

Similarly, assuming (15) for $k-1$,

$$\xi_{n+k-1}' \le 9(n-k+2)$$

so that

$$\xi_{n+k}' = \left[\frac{\xi_{n+k-1}'+\xi_{n+k}}{10}\right] \le \left[\frac{(9(n-k+2)+(n-k+1)9^2}{10}\right],$$

since by (13) $\xi_{n+k} \le (n-k+1)9^2$.

Hence

$$\xi_{n+k}' \le \left[\frac{(n-k+1)90+9}{10}\right] = 9(n-k+1)$$

which completes the induction on $k$ and upholds (15).

In particular, for $k=n$ in (15), we get

$$\xi_{2n}' \le 9$$

which is what we wanted to show.

Accordingly, all the coefficients $a_0, a_1, \ldots, a_{2n-1}, a_{2n}$ satisfy the inequalities (12).

It remains to verify that

(16) $$X^2 = a_0 + a_1.10 + a_2.10^2 + \ldots + a_{2n}.10^{2n}$$

By the definition of $\xi'_k (k = 0, 1, \ldots, 2n - 1)$ we have:

(i) Divide $\xi_0$ by 10 to get the quotient $\xi'_0$ and the remainder $a_0$

(ii) Divide $\xi'_{k-1} + \xi_k$ by 10 to get the quotient $\xi'_k$ and the remainder $a_k$

Hence

$$
\begin{aligned}
\xi_0 &= 10\xi'_0 + a_0 \\
\xi'_0 + \xi_1 &= 10\xi'_1 + a_1 \\
&\vdots \\
\xi'_{2n-2} + \xi_{2n-1} &= 10\xi'_{2n-1} + a_{2n-1} \\
\xi'_{2n-1} + \xi_{2n} &= 10\xi'_{2n} + a_{2n} \\
\xi'_{2n} &= a_{2n+1}
\end{aligned}
\tag{17}
$$

In (17) we multiply the second equation by 10, the third by $10^2, \ldots$ and the last by $10^{2n+1}$. The terms involving

$$\xi'_0, \xi'_1, \ldots, \xi'_{2n}$$

cancel and we find

$$\xi_0 + \xi_1.10 + \ldots + \xi_{2n}.10^{2n} = a_0 = a_1.10 + \ldots + a_{2n}.10^{2n} + a_{2n+1}.10^{2n+1}$$

since the left side here is $X^2$ we have proved (16).

Thereby, the theorem is completely established. ■

**Example 1.** To find the square of $X = 987654321$

Step 1: To compute the duplexes $\xi_k$

Step 2: $a_0 =$ the last digit of $\xi_0$, $\xi'_0$ is obtained from $\xi_0$ by removing its last digit $a_0$.

$a_k =$ the last digit of $\xi'_{k-1} + \xi_k$, and

$\xi'_k$ is obtained from $\xi'_{k-1} + \xi_k$ by removing its last digit $a_k$ $(k = 0, 1, \ldots, 2n)$

$\xi'_{2n} = a_{2n+1}$

We can tabulate the results as follows:

| | | | | | | | | | | | | | | | | | | | |
|---|---|---|---|---|---|---|---|---|---|---|---|---|---|---|---|---|---|---|---|
| | | 81 | 144 | 190 | 220 | 235 | 236 | 224 | 200 | 165 | 120 | 84 | 56 | 35 | 20 | 10 | 4 | 1 | $\xi_k$ |
| | | 9 | 16 | 21 | 24 | 26 | 26 | 24 | 21 | 17 | 12 | 8 | 5 | 3 | 2 | 1 | 0 | 0 | $\xi'_k$ |
| $X^2$ | 9 | 7 | 5 | 4 | 6 | 1 | 0 | 5 | 7 | 7 | 8 | 9 | 9 | 7 | 1 | 0 | 4 | 1 | $a_k$ |

The only computation is in finding the duplexes:

$$\begin{cases} \xi_0 & = 10\xi'_0 + a_0 \\ \xi'_{k-1} + \xi_k & = 10\xi'_k + a_k \qquad (k = 1, \ldots 2n) \\ \xi'_{2n} & = a_{2n+1} \end{cases}$$

**Example 2.** To find the square of $X = 8978$

| | | | | | | | | |
|---|---|---|---|---|---|---|---|---|
| | 64 | 144 | 193 | 254 | 193 | 112 | 64 | $\xi_k$ |
| | 8 | 16 | 22 | 27 | 20 | 11 | 6 | $\xi'_k$ |
| $X^2 = 8$ | 0 | 6 | 0 | 4 | 4 | 8 | 4 | $a_k$ |

**Example 3.** To find the square of $X = 7654321$

| | | | | | | | | | | | | | |
|---|---|---|---|---|---|---|---|---|---|---|---|---|---|
| | 49 | 84 | 106 | 116 | 115 | 104 | 84 | 56 | 35 | 20 | 10 | 4 | 1 | $\xi_k$ |
| | 5 | 9 | 11 | 12 | 12 | 11 | 8 | 5 | 3 | 2 | 1 | 0 | 0 | $\xi'_k$ |
| $X^2 = 5$ | 8 | 5 | 8 | 8 | 6 | 2 | 9 | 9 | 7 | 1 | 0 | 4 | 1 | $a_k$ |

## § 4. SQUARE ROOT

We shall describe ARYABHATA'S method of finding the square root:

भाग हरे दवर्गान्नित्यं द्विगुणेन वर्गमूलेन ।
वर्गाद्वर्गे शुध्दे लब्धं स्थानान्तरे मूलम् ।।

Given a number mark the *even* ( ) places by horizontal lines and the *odd* ( ) places by vertical lines, thus

```
| − | − |         | − | − |
1 1 9 7 1 6,      5 4 7 5 6
```

$1^0$ Subtract the greatest possible square from the last odd place;

$2^0$ Always divide the even place by twice the square root upto the preceding odd place;

$3^0$ Subtract from the odd place (standing on the right) the square of the quotient;

$4^0$ Repeat the process as long as there are still digits on the right..

**Example 1.** To find the square root of $A = 119716$

We have $9 < 11 < 16$, $3 < \sqrt{11} < 4$, so that the integral part of $\sqrt{11}$ is = 3

$$\begin{array}{llll}
 & & 1\ 1\ \ 9\ \ 7\ 1\ 6\ (\mathbf{3} & \\
 & & -\ 3^2 & \\
\text{Divide by } 2\times 3 & 6) & 2\ \ 9\ \ (\ \mathbf{4} & \\
 & & 2\ \ 4 & \\
 & & \quad 5\ \ 7 & \\
 & & \quad -\ 4^2 & \sqrt{A} = 346 \\
\text{Divide by } 2\times 34 & 68) & \quad 4\ \ 1\ \ 1\ \ (\ \mathbf{6} & \\
 & & \quad 4\ \ 0\ \ 8 & \\
 & & \qquad 3\ \ 6 & \\
 & & \qquad -\ 6^2 & \\
 & & \qquad\ * &
\end{array}$$

This can be proved as follows:

(i) put $x_1 = [\sqrt{11}]$, $\underline{x_1 = 3}$
$11 - x_1^2 = 2$

(ii) Divide 29 by $2x_1$ with quotient $x_2$, $\underline{x_2 = 4}$
$29 = 2x_1x_2 + 5$

(iii) $57 - x_2^2 = 41$

(iv) Divide 411 by $2(10x_1 + x_2) = 2 \times 34$
$411 = 2(10x_1 + x_2).x_3 + 3$, $\underline{x_3 = 6}$

(v) $36 - x_3^2 = 0$

The process ends, and the given number is a perfect square.

Thus we have

$$\begin{array}{|r}
11 = x_1^2 + 2 \\
29 = 2x_1x_2 + 5 \\
57 = x_2^2 + 41 \\
411 = 2x_3(10x_1 + x_2) + 3 \\
36 = x_3^2
\end{array}$$

Multiply these equations, in order, by $10^4$, $10^3$, $10^2$, 10, 1 and add. Observing that

$$2.10^4 = 20.10^3, \quad 5.10^3 = 50.10^2, \quad 41.10^2 = 410.10$$

and cancelling the corresponding terms we obtain

$$11.10^4 + 9.10^3 + 7.10^2 + 1.10 + 6$$
$$= x_1^2.10^4 + 2x_1x_2.10^3 + x_2^2.10^2 + 2x_1x_3.10 + x_3^2$$

in other words

$$\begin{aligned} 119716 &= (x_1.10^2 + x_2.10 + x_3)^2 \\ &= (3.10^2 + 4.10 + 6)^2 \\ &= (346)^2 \end{aligned}$$

Consequently

$$\sqrt{119716} = 346$$

**Example 2.** To find the square root of $A = 738915489$ we proceed according to the rule of Aryabhata:

```
          |  —  |  —  |  —  |  —  |
          7  3  8  9  1  5  4  8  9
        - 2^2
  4)      3  3  7  ( 7                 2
          2  8
          ----
             5  8
           - 7^2                       7
          ----
 54)         9  9  ( 1
             5  4
             -------
             4  5  1
              - 1^2                    1
          -----------
542)      4  5  0  5  ( 8
          4  3  3  6
          ----------
          1  6  9  4
             - 8^2                     8
          -------------
5436)     1  6  3  0  8  ( 3
          1  6  3  0  8
          -------------
                0  9
             - 3^2  (                  3
          -------------
                *
```

When we divided 33 by 4 we could have taken the quotient as 8 and the remainder as 1; but then, at the next stage we will meet with

$$\begin{array}{r} 18 \\ -8^2 \\ \hline \end{array}$$

Such a situation is always to be avoided.

$$\sqrt{738915489} = 27183$$

Shri Bharathi Krishna Thirthaji has a variant which has much elegance and symmetry. It is based on the formula for the square of an integer in duplex notation which we have treated in the previous section.

We have

$$[x_1+x_2]^2 = (x_1) + (x_1x_2) + (x_2)$$

$$[x_1+x_2+x_3]^2 = (x_1) + (x_1x_2) + (x_1x_2x_3) + (x_2x_3) + (x_3)$$

$$[x_1+x_2+x_3+x_4]^2 = (x_1) + (x_1x_2) + (x_1x_2x_3) + (x_1x_2x_3x_4) + (x_2x_3x_4) + (x_3x_4) + (x_4)$$

In general, we have

(1)

$$[x_1+x_2+\ldots+x_n]^2 = \sum_{k=1}^{n}(x_1,x_2\ldots x_k) + \sum_{k=2}^{n}(x_kx_{K+1}\ldots x_n)$$

**Theorem 1. [Shri Bharathi Krishna Thirthaji]**

*Given:* a positive integer

$$A = a_0.10^8 + a_1.10^7 + a_2.10^6 + a_3.10^5 + a_4.10^4 + a_5.10^3 + + a_6.10^2 + a_7.10 + a_8$$

$$0 \le a_k \le 9 \qquad \text{for } k = 1, 2, \ldots, 8$$

$$0 < a_0 < 10^2$$

$1^0$ Define the numbers (positive integers):

$$r_1, r_2, \ldots, r_5 \text{ and } x_0, x_1, \ldots, x_4$$

by the chain of equations

(2)
$$\begin{array}{|rl} a_0 & = (x_0) + r_1 \\ 10r_1 + a_1 & = (x_0x_1) + r_2 \\ 10r_2 + a_2 & = (x_0x_1x_2) + r_3 \\ 10r_3 + a_3 & = (x_0x_1x_2x_3) + r_4 \\ 10r_4 + a_4 & = (x_0x_1x_2x_3x_4) + r_5 \end{array}$$

$2^0$ Define the number $r_6, r_7, r_8, r_9$ from the equations

(3)
$$\begin{array}{|rl} 10r_5 + a_5 & = (x_1, x_2, x_3, x_4) + r_6 \\ 10r_6 + a_6 & = (x_2, x_3, x_4) + r_7 \\ 10r_7 + a_7 & = (x_3, x_4) + r_8 \\ 10r_8 + a_8 & = (x_4) + r_9 \end{array}$$

$3^0$ Put

$$X = x_0.10^4 + x_1.10^3 + x_2.10^2 + x_3.10^3 + x_4$$

*Claim*:

$$\boxed{A = X^2 \text{ if and only if } r_9 = 0}$$

*Remark*: Given a number $T$ we can write

$$T = 10T' + t_0 \quad (0 \leq t_0 \leq 9)$$

$t_0$ = the last digit of $T$, and $T'$ results from $T$ by removing from it its last digit. We have already mentioned this before(See p. 78). Hence in (2) and (3) the numbers

$$10r_k + a_k \quad (k = 1, \ldots 8)$$

have $a_k$ as the last digit with the digits of $r_k$ as the remaining ones.

**Proof**: The proof rests on the formula (1) *[See p.85]*. Multiply the chain of equations (2), in order, by

$$10^8, 10^7, 10^6, 10^5, 10^4$$

and the chain of equations (3), in order, by

$$10^3, 10^2, 10, 1$$

and add. The terms involving

$$r_1, r_2, \ldots, r_8$$

all cancel. On the left side we obtain

$$a_0.10^8 + a_1.10^7 + a_2.10^6 + \ldots + a_7.10 + a_8$$

i.e. the number $A$, and on the right side we get

$$(x_0).10^8 + (x_0x_1).10^7 + (x_0x_1x_2).10^6 + (x_0x_1x_2x_3).10^5 +$$
$$(x_0x_1x_2x_3x_4).10^4 + (x_1x_2x_3x_4).10^3 + (x_2x_3x_4).10^2 +$$
$$(x_3x_4).10 + (x_4)$$

which is See (4), p. 76 as also (1) p. 85

$$[x_0.10^4 + x_1.10^3 + x_2.10^2 + x_3.10 + x_4]^2$$

Consequently, we have the identity

$$A = X^2 + r_9$$

from which follows the claim of the theorem ■

The computations can be arranged according to the scheme

| $a_0$ | $a_1$ | $a_2$ | $a_3$ | $a_4$ | $a_5$ | $a_6$ | $a_7$ | $a_8$ |
|---|---|---|---|---|---|---|---|---|
| ↑ | $r_1$ | $r_2$ | $r_3$ | $r_4$ | $r_5$ | $r_6$ | $r_7$ | $r_8$ |
| $x_0$ | $x_1$ | $x_2$ | $x_3$ | $x_4$ | | | | |

Since $x_0, x_1, \ldots x_4$ are already determined by (2) the square root $X$ is determined by (2) alone (provided A is a perfect square). We use (3) to verify that $r_9 = 0$ and ascertain ourselves that A is indeed a perfect square (or conclude that it is not so when $r_9 \neq 0$).

**Example 2.** [Continued]

$$\underline{A = 738915489}$$

In order to find the square root $X = \sqrt{A} = x_0.10^4 + x_1.10^3 + x_2.10^2 + x_3.10 + x_4$ we have to solve the system of equations:

$$a_0 = 7 = (x_0) + r_1, \quad x_0 = [\sqrt{7}] = 2, \quad r_1 = 3$$
$$10r_1 + a_1 = 33 = (x_0x_1) + r_2, \quad x_1 = 7, r_2 = 5$$
$$10r_2 + a_2 = 58 = (x_0x_1x_2) + r_3, \quad x_2 = 1, r_3 = 5$$
$$10r_3 + a_3 = 59 = (x_0x_1x_2x_3) + r_4, \quad x_3 = 8, r_4 = 13$$
$$10r_4 + a_4 = 131 = (x_0x_1x_2x_3x_4) + r_5, \quad x_4 = 3, r_5 = 6$$

Then $r_6, r_7, r_8, r_9$ must be found from

(5)
$$10r_5 + a_5 = 65 = (7, 1, 8, 3) + r_6 \qquad r_6 = 7$$
$$10r_6 + a_6 = 74 = (1, 8, 3) + r_7 \qquad r_7 = 4$$
$$10r_7 + a_7 = 48 = (8, 3) + r_8 \qquad r_8 = 0$$
$$10r_8 + a_8 = 09 = (3) + r_9 \qquad \boxed{r_9 = 0}$$

Consequently $A = X^2$, and $X = 27183$

*Remark*: In Theorem 1 §3. (p.79) we had used the notation

$$X = x_0 + x_1.10 + \ldots + x_n.10^n$$

so that

$$X = \langle x_0, x_1, \ldots, x_{n-1}, x_n \rangle_{10}$$

but in the proof of Theorem 1 of the present section we have found it convenient to take $X$ as

| | | | | | | | | | | |
|---|---|---|---|---|---|---|---|---|---|---|
| | $a_k$ | 7 | 3 | 8 | 9 | 1 | 5 | 4 | 8 | 9 |
| $(10r_k + a_k)\uparrow$ | $r_k$ | | 3 | 5 | 5 | 13 | 6 | 7 | 4 | 0 |
| | $x_k$ | 2 | 7 | 1 | 8 | 3 | | | | |

The duplexes for $X$ are obtained from (4), (5):

(6)
$$(3) = 9 \qquad (2718) = 46$$
$$(83) = 48 \qquad (271) = 53$$
$$(183) = 70 \qquad (27) = 28$$
$$(7183) = 58 \qquad (2) = 4$$
$$(27183) = 125$$

We can make a final verification that $A = X^2$ by computing $X^2$ using the information (6)

| | | | | | | | | | | |
|---|---|---|---|---|---|---|---|---|---|---|
| | | 4 | 28 | 53 | 46 | 125 | 58 | 70 | 48 | 9 | |
| $(10r_k + a_k)\downarrow$ | | 3 | 5 | 5 | 13 | 6 | 7 | 4 | 0 | $r_k$ |
| | 7 | 3 | 8 | 9 | 1 | 5 | 4 | 8 | 9 | $a_k$ |

**Example 3.** To find the square root of $A = 45319824$. A has an even number of digits. Pairing two digits each time, starting from the right, we arrive at the last group 45.

Put

$$a_0 = [\sqrt{45}], \quad \text{then } a_0 = 6$$

We find

$$\begin{array}{r|l} 10^8 & 45 = (6) + 9 \\ 10^5 & 93 = (6,7) + 9 \\ 10^4 & 91 = (6,7,3) + 6 \\ \text{Multiply by } 10^3 & 69 = (6,7,3,2) + 3 \\ 10^2 & 38 = (7,3,2) + 3 \\ 10 & 12 = (3,2) \\ 1 & 04 = (2) \end{array}$$

Adding we get

$$A = [6.10^3 + 7.10^2 + 3.10 + 2]^2 = (6732)^2$$

## § 5. CUBE

SRIDHARA (750 AD) says:

Set down the cube of the last; then set down at the next place the square of the last multiplied by thrice the succeeding; then at the next place the product of the square of the succeeding and last multiplied by three; finally put at the next place the cube of the succeeding.

BHASKARA (1150 AD) says:

In order to find the cube of a given number set down the cube of the *last* digit; then the square of the last multiplied by three times the succeeding multiplied by three times the last and then the cube of the succeeding: these placed so that there is the difference of a place between one result and the next, and added give the cube.

**Example 1.** To find the cube of 12

$12 = 10a_1 + 2a_0, (a = 2, a_1 = 1), (a_1$ is the *last* digit)

$$\begin{array}{rl|r} a_1^3 & = 1 & 1 \\ 3a_1^2 a_0 & = 6 & .6 \\ 3a_1 a_0^2 & = 12 & 12 \\ a_0^3 & = 8 & .8 \\ \hline & & 12^3 = 1728 \end{array}$$

**Example 2.** To find $(123)^3$

$123 = 10a + b, (a = 12, b = 3)$

$$\begin{array}{rl|r} a^3 & = 12^3 = 1728 & 1\,7\,2\,8\,.\,.\,. \\ 3a^2 b & = 1296 & 1\,2\,9\,6\,.\,. \\ 3ab^2 & = 324 & 3\,2\,4\,. \\ b^3 & = 27 & 2\,7 \\ \hline & & 123^3 = 1\,8\,6\,0\,8\,6\,7 \end{array}$$

**Example 3:** To find $(1234)^3$

Put $c = 123$ and $d = 4$, then $1234 = 10c + d$, $(1234)^3 = (10c + d)^3 = c^3.10^3 + (3c^2d).10^2 + (3cd^2)10 + d^3$

We have

$$c^2 = (123)^2 = 15129 \qquad \begin{array}{ccccc} \hline 1 & 4 & 10 & 12 & 9 \\ & 0 & 1 & 1 & 0 \\ \hline 1 & 5 & 1 & .2 & 9 \end{array}$$

By Example 2,

$$c^3 = (123)^3 = 1860867$$

$$3c^2 d = 181548, 3cd^2 = 5904$$

Therefore we have

$$(1234)^3 = \begin{array}{|r} 1\,8\,6\,0\,8\,6\,7\,.\,.\,. \\ 1\,8\,1\,5\,4\,8\,.\,. \\ 5\,9\,0\,4\,. \\ 6\,4 \\ \hline 1\,8\,7\,9\,0\,8\,0\,9\,0\,4 \end{array}$$

**Example 4.** To find $(10000012)^3$

We have $10000012 = (10^5 + 12)^3 = (10^5 + a)^3$ with $a = 12$

Thus we have to find

$$10^{15} + 3.10^{10}.a + 3.10^5.a^2 + a^3$$

Since $3a = 36$ and $3a^2 = 432$,

$$(10000012)^3 = \begin{array}{|l} 1\ .\ .\ .\ .\ .\ \ .\ .\ .\ .\ .\ \ .\ .\ .\ .\ . \\ \qquad\quad 3\,6\ \ .\ .\ .\ .\ .\ \ .\ .\ .\ .\ . \\ \qquad\qquad\qquad 4\,3\,2\ \ .\ .\ .\ .\ . \\ \qquad\qquad\qquad\qquad\quad 1\,7\,2\,8 \\ \hline 1\ 0\,0\,0\,3\,6\ \ 0\,0\,4\,3\,2\ \ 0\,1\,7\,2\,8 \\ \hline \end{array}$$

**Example 5.** [Shridhara (750 AD)]

Show that

$$\boxed{\sum_{k=1}^{n}\{3k(k-1)+1\} = n^3}$$

We have the identity

$$(k+1)^3 = k^3 + 3k^2 + 3k + 1$$

so that

$$(k+1)^3 - k^3 = 3k(k+1) + 1$$

Setting $k = 0, 1, 2, \ldots, n-1$ and adding

$$\begin{aligned} 1^3 - 0 &= 1 \\ 2^3 - 1^3 &= 3.(1.2) + 1 \\ 3^3 - 2^3 &= 3(2.3) + 1 \\ &\vdots \\ n^3 - (n-1)^3 &= 3.(n-1).n + 1 \end{aligned}$$

we get

$$n^3 = 3[1.2 + 2.3 + \ldots + (n-1)n] + n$$

which yields the required result.

## § 6. CUBE ROOT

Aryabhata (I) (499 A.D.) gives the following rule for finding the cube root of a number.

Let us call the units place "ghana" the 10-place "the first Aghana", the $10^2$ place the "second Aghana" : We take in groups of *three* digits starting from the right.

| Second Aghana | First Aghana | Units |
|---|---|---|
| $10^2$ | $10$ | 1 |
| $10^4$ | $10^3$ | |
| $10^6$ | $10^5$ | |
| ⋮ | ⋮ | |

$1^0$ Subtract the greatest possible cube from the last ghana place.

$2^0$ Divide the second aghana place (standing on the right of the last ghana place) by thrice the square of the cube root already obtained in $1^0$

$3^0$ Subtract from the first Aghana place (standing on the right of the second Aghana place) the square of the quotient multiplied by the previous cube root.

$4^0$ Subtract the cube of the quotient from the ghana place (standing on the right of the first aghana place)

$5^0$ Repeat the process until all the digits are exhausted.

**Example 1.** Find the cube root of 1 953 125

$$\begin{array}{llr}
 & \overset{|}{1}\ \overset{-}{9}\ \overset{-}{5}\ \overset{|}{3}\ \overset{-}{1}\ \overset{-}{2}\overset{|}{5} & \\
 & -1^3 & \mathbf{1} \\
\hline
3.1^2 = 3) & 0\ 9\ \ (\ 2 & \\
 & \ \ \ 6 & \\
\hline
 & \ \ \ 3\ 5 & \\
 & -3.\ 2^2.\ 1 & \\
\hline
 & \ \ \ 2\ 3\ 3 & \\
 & \ \ -\ 2^3 & \mathbf{2} \\
\hline
3.12^2 = 432) & 2\ 2\ 5\ 1\ \ (\ 5 & \\
 & 2\ 1\ 6\ 0 & \\
\hline
 & \ \ 9\ 1\ 2 & \\
 & -\ 3.\ 5^2\ .1\ 2 & \\
\hline
 & \ \ \ \ 1\ 2\ 5 & \mathbf{5} \\
 & \ \ \ \ \ -\ 5^3 & \\
\hline
\end{array}$$

The process ends and the cube root is 125: $\sqrt[3]{1953125} = 125$

**Example 2.** Find the cube root of $A = 76\ 928\ 302\ 277$

$$\begin{array}{rl}
 & \overset{|}{7}\ \overset{-}{6}\ \ \overset{-}{9}\ \overset{|}{2}\ \overset{-}{8}\ \overset{-}{3}\ \ \overset{|}{0}\ \overset{-}{2}\ \overset{-}{2}\ \overset{|}{7}\ 7 \\
 & -\,4^3 \qquad\qquad\qquad\qquad\qquad 4^3 < 76 < 5^3 \\
3.4^2 = 48\,) & 1\ 2\ 9\ (\ 2 \qquad\qquad \mathbf{4} \qquad [\sqrt[3]{76}] = 4 \\
 & \ \ 9\ 6 \\
 & 3\ \ 3\ 2 \qquad\qquad\qquad \mathbf{2} \\
 & -\,3\cdot\ 2^2\cdot 4 \qquad\qquad \mathbf{5} \\
 & 2\ \ 8\ 4\ 8 \\
 & \qquad -\,2^3 \qquad\qquad \mathbf{3} \\
3.42^2 = 5292) & 2\ \ 8\ 4\ 0\ 3\ \ (\ 5 \qquad \sqrt[3]{A} = 4253 \\
 & 2\ \ 6\ 4\ 6\ 0 \\
 & 1\ \ 9\ 4\ 3\ 0 \\
 & -\,3\cdot\ 5^2\cdot\ (\ 4\ 2\ ) \\
 & 1\ \ 6\ 2\ 8\ 0\ 2 \\
 & \qquad\qquad -\,5^3 \\
3.(425)^2 = 541875) & 1\ \ 6\ 2\ 6\ 7\ 7\ 2\ (\ 3 \\
 & 1\ \ 6\ 2\ 5\ 6\ 2\ 5 \\
 & \quad 1\ 1\ 4\ 7\ 7 \\
 & \quad -\,3\cdot\ 3^2\cdot 4\ 2\ 5 \\
 & \qquad\qquad 2\ 7 \\
 & \qquad\qquad -\,3^3 \\
 & \qquad\qquad\quad * \qquad\qquad \sqrt[3]{A} = 4253
\end{array}$$

We shall give a proof of Aryabhata's rule for determining the cube root.

**Theorem 1.** [Aryabhata]

*Given a positive integer*

(1) $$A = a_1.10^3 + a_2.10^2 + a_3.10 + a_4$$

$$0 < a_1 < 10^3, 0 \leq a_2, a_3, a_4 \leq 9 \ integers$$

*We put*

(2) $$x_1 = [\sqrt{a_1}], \quad b_1 = a_1 - x_1^3$$

*Then there exists an integer* $x_2, 0 \leq x_2 \leq 9$, *such that*

(3) $A = [10x_1 + x_2]^3 + A'$

**Proof:** $1^0$ We have $x_1 \leq \sqrt[3]{a_1} < x + 1$ so that $x_1^3 \leq a_1 < (x_1 + 1)^3$. We remove the largest possible cube from $a_1$. Since $0 < a_1 < 10^3$, we must have $x_1 \leq \sqrt[3]{a_1} < 10$, i.e., $1 \leq x_1 \leq 9$. From (2):

(4) $$a_1 = x_1^3 + b_1$$

$2^0$ Divide $10b_1 + a_2$ by $3x_1^2$ to get the quotient $x_2$ and remainder $b_2$:

(5) $$10b_1 + a_2 = 3x_1^2.x_2 + b_2$$

$3^0$ Subtract $3x_1.x_2^2$ from $10b_2 + a_3$ to get $b_3$ :

(6) $$10b_2 + a_3 = 3x_1.x_2^2 + b_3$$

$4^0$ Subtract $x_2^3$ from $10b_3 + a_4$ to get $b_4$ :

(7) $$10b_3 + a_4 = x_2^3 + b_4$$

From (4), (5), (6), (7) we have

(8) $$\left| \begin{array}{ll} a_1 & = x_1^3 + b_1 \\ 10b_1 + a_2 & = 3x_1^2.x_2 + b_2 \\ 10b_2 + a_3 & = 3x_1.x_2^2 + b_3 \\ 10b_3 + a_4 & = x_2^3 + b_4 \end{array} \right.$$

Multiplying the equations (8), in order, by $10^3$, $10^2$, 10 and 1 and adding we get, [the terms containing $b_1$, $b_2$, $b_3$ cancel],

$$\begin{aligned} a_1.10^3 + a_2.10^2 + a_3.10 + a_4 &= x_1.10^3 + 3(x_1.10)^2 x_2 \\ &\quad + 3(x_1.10)x_2^2 + x_2^3 + b_4 \\ &= [x_1.10 + x_2]^3 + b_4 \end{aligned}$$

which yields the assertion (3) of the theorem with $A' = b_4$ and completes the proof.

*Remarks*:

Since in (1) we have assumed $1 \leq a_1 \leq 10^3$, $a$ has 1,2, or 3 digits so that A is a number with utmost 6 digits. It is important to remember that the system of equations (8) must be solved for *non-negative integers*

$$x_1, x_2 \text{ and } b_1, b_2, b_3, b_4$$

In the second equation of (8) one usually divides $10b_1 + a_2$ by $3x_1^2$ and gets the quotient $x_2$ and the remainder $b_2$, in other words: one removes the *largest* possible multiple $x_2$ of $3x_1^2$, i.e.

(9)
$$x_2 = \left[\frac{10b_1 + a_2}{3x_1^2}\right]$$

(where [...] is the integral part). However it may turn out that $b_2$ may be too small to satisfy the third equation in (8) because it is requried that

(10)
$$10b_2 + a_3 > 3x_1x_2^2$$

In such an event we have to choose the next lower multiple of $3x_1^2$: [namely $x_2 - 1$ instead of $x_2$ in (9)]

**Theorem 2.** [Aryabhata]

*Given a positive integer*

(11)
$$A = a_1.10^9 + a_2.10^8 + \ldots + a_9.10 + a_{10}$$
$$1 \le a_1 \le 10^3, 0 \le a_k \le 9 \quad (k = 2, 3, \ldots 10)$$

Solve the following system of equations for *integers* $x_1$, $x_2$, $x_3$, $x_4$, such that

$$1 \le x_1 \le 9, \quad 0 \le x_2, \quad x_3, \quad x_4 \le 9$$

$I^0$
$$a_1 = r_1^3 + b_1$$
$$10b_1 + a_2 = 3r_1^2x_2 + b_2 \quad \Big| \quad r_1 = x_1$$
$$10b_2 + a_3 = 3r_1x_2^2 + b_3 \quad \Big| \quad A_1 = a_1.10^3 + a_2.10^2 + a_3.10 + a_4$$
$$10b_3 + a_4 = x_2^3 + b_4$$

$II^0$
$$A_1 = r_2^3 + b_4$$
$$10b_4 + a_5 = 3r_2^2x_3 + b_5 \quad \Big| \quad r_2 = 10r_1 + x_2$$

(13)
$$10b_5 + a_6 = 3r_2x_3^2 + b_6 \quad \Big| \quad A_2 = a_5.10^2 + a_6.10 + a_7$$
$$10b_6 + a_7 = x_3^3 + b_7$$

$III^0$
$$A_1.10^3 + A_2 = r_3^3 + b_7$$
$$10b_7 + a_8 = 3r_3^2x_4 + b_8 \quad \Big| \quad r_3 = 10r_2 + x_3$$
$$10b_8 + a_9 = 3r_3x_4^2 + b_9 \quad \Big| \quad A_3 = a_8.10^2 + a_9.10 + a_{10},$$

$$10b_9 + a_10 = x_4^3 + b_10$$

*Claim:*

(14) $$\boxed{A = [x_1.10^3 + x_2.10^2 + x_3.10 + x_4]^3 + b_{10}}$$

Consequently

$$\boxed{A \text{ is a perfect cube } \Longleftrightarrow b_{10} = 0}$$

**Proof:** Applying Theorem 1 to the group $I^0$ of equations we get

$$A_1 = [10r_1 + x_2]^3 + b_4 = r_2^3 + b_4$$

Then, by the group $II^0$ of equations

$$A_1.10^3 + a_5.10^2 + a_6.10 + a_7 = [10r_2 + x_3]^3 + b_7 = r_3^3 + b_7$$

that is:

$$A_1.10^3 + A_2 = r_3^3 + b_7$$

Finally by the system $III^0$

$$\begin{aligned} &(A_1.10^3 + A_2).10^3 + a_8.10^2 + a_8.10^2 + a_9.10 \\ &= (10r_3 + x_4)^3 + b_{10} \\ &= (10^2r_2 + 10x_3 + x_4)^3 + b_10 \\ &= [x_1.10^3 + x_2.10^2 + x_3.10 + x_4]^3 + b_{10} \end{aligned}$$

which establishes the claim and completes the proof. The computations needed for finding the cube root may be conveniently tabulated as follows:

$$\begin{array}{r|l|} r_1 = x_1 = & a_1 \\ = [\sqrt[3]{a_1}] & -x_1^3 \\ \hline & b_1.10 + a_2 \\ & -3x_1^2.x_2 \\ \hline \text{I}^\circ & b_2.10 + a_3 \\ & -3x_1x_2^2 \\ \hline & b_3.10 + a_4 \\ & -x_2^3 \\ \hline & b_4 \end{array} \qquad \begin{array}{r|l|} \underline{r_2 = 10r_1 + x_2} & b_4.10 + a_5 \\ & -3r_2^2.x_3 \\ \hline & b_5.10 + a_6 \\ & -3r_2x_3^2 \\ \hline \text{II}^\circ & b_6.10 + a_7 \\ & -r_2^3 \\ \hline & b_7 \end{array}$$

$$
\begin{array}{c|l|l}
\underline{r_3 = 10r_2 + x_3} & b_7.10 + a_8 & \begin{cases} A_1 & = r_2^3 + b_4 \\ A_1.10^3 + A_2 & = r_3^3 + b_7 \\ A_1.10^6 + & \\ +A_2.10^3 + A_3 & = r_4^3 + b_{10} \end{cases} \\
 & -3r_3^2.x_4 & \\
\hline
 & b_8.10 + a_9 & \begin{cases} r_2 & = 10x_1 + x_2 \\ r_3 & = 10^2x_1 + 10x_2 + x_3 \\ r_4 & = 10^3x_1 + 10^2x_2 + 10x_3 + x_4 \end{cases} \\
 & -3r_3x_2^4 & \\
\hline
\text{III}^\circ & b_9.10 + a_{10} & \\
 & -x_4^3 & \\
\hline
 & b_{10} &
\end{array}
$$

$$A_1 = \langle a_1, a_2, a_3, a_4 \rangle$$

$$A_1.10^3 + A_2 = \langle a_1, a_2, a_3, a_4, a_5, a_6, a_7 \rangle$$

$$A_1.10^6 + A_2.10^3 + A_3 = \langle a_1, a_2, a_3, a_4, a_5, a_6, a_7, a_8, a_9, a_{10} \rangle = A$$

$$A = [x_1.10^3 + x_2.10^2 + x_3.10 + x_4]^3 + b_{10}$$

**Example 3.**

To find the cube root of $A = 792\ 994\ 249\ 216$. In the notation of Theorem 2, we have

$$A_1 = 792\ 994$$

$$A_1.10^3 + A_2 = 792\ 994\ 249$$

$$A_1.10^6 + A_2.10^3 + A_3 = 792\ 994\ 249\ 216$$

The computations of the three groups of equations in (13) arc tabulated as follows:

| $k$ | 1 | 2 | 3 | 4 | 5 | 6 | 7 | 8 | 9 | 10 |
|---|---|---|---|---|---|---|---|---|---|---|
| $a_k$ | 792 | 9 | 9 | 4 | 2 | 4 | 9 | 2 | 1 | 6 |
| $b_k$ | 63 | 153 | 1431 | 14306 | 16102 | 154124 | 1541124 | 9992 | 21 | 0 |
| $x_k$ | 9 | 2 | 5 | 6 | | | | | | |
| $r_k$ | 9 | 92 | 925 | 9256 | | | | | | |

$\sqrt[3]{A} = \mathbf{9256}$

| | | |
|---|---|---|
| | 792 | **9** |
| | $-9^3$ | |
| $3.9^2$ | 639 | **2** |
| | $-3.9^2.2$ | |
| | 1539 | |
| | $-3.9.2^2$ | |
| | 14314 | |
| | $-2^3$ | |
| $b_4$ | 14306 | |

| | | |
|---|---|---|
| $3.92^2$ | 143062 | **5** |
| | $-3.92^2.5$ | |
| | 161024 | |
| | $-3.92.5^2$ | |
| | 1541249 | |
| | $-5^3$ | |
| $b_7 =$ | 1541124 | |

| | | |
|---|---|---|
| $3.925^2$ | 15411242 | **6** |
| | $-3.925^2.6$ | |
| | 9992.1 | |
| | $-3.925.6^2$ | |
| | 216 | |
| | $-6^3$ | |
| $b_{10} =$ | 0 | |

We shall now indicate how congruences can be used to find the digits of the cube root. If

$$A = X^3, X = x_1.10^3 + x_2.10^2 + x_3.10 + x_4$$

we have

$$X \equiv x_4(\text{mod } 10), \quad X \equiv (x_3.10 + x_4)(\text{mod } 10^2),$$
$$X \equiv x_2.10^2 + x_3.10 + x_4(\text{mod } 10^3)$$

whence follows:

$$(15) \qquad A \equiv X^3 \equiv \left| \begin{array}{l} x_4^3 (\text{mod } 10) \\ (x_3.10 + x_4)^3 (\text{mod } 10^2) \\ (x_2 \cdot 10^2 + x_3 \cdot 10 + x_4)^3 (\text{mod } 10^3) \end{array} \right.$$

If $A = a_1.10^9 + a_2.10^8 + \ldots + a_8.10^2 + a_9.10 + a_{10}$, we obtain

$$(16) \qquad A \equiv \left| \begin{array}{l} a_{10} (\text{mod } 10) \\ a_9.10 + a_{10}(\text{mod } 10^2) \\ a_8.10^2 + a_9.10 + a_{10}(\text{mod } 10^3) \end{array} \right.$$

From (15) and (16) we get congruences for solving $x_1, \quad x_2, \quad x_3, \quad x_4$

**Example 4.** Find the cube root of $A = 927\ 715\ 200\ 777$

Let $A = X^3$, with $X = x_1.10^3 + x_1.10^2 + x_3.10 + x_4$

Recall the cubes of $1, 2, \ldots 9$

| | | | | | | | | | |
|---|---|---|---|---|---|---|---|---|---|
| | 1 | 8 | 7 | 4 | 5 | 6 | 3 | 2 | 9 |
| (17) | $1^3$ | $2^3$ | $3^3$ | $4^3$ | $5^3$ | $6^3$ | $7^3$ | $8^3$ | $9^3$ |
| | 1 | 8 | 27 | 64 | 125 | 216 | 343 | 512 | 729 |

Since

$$A \equiv 7(\text{mod } 10), X \equiv x_4(\text{mod } 10), A = X^3$$

we have (See (16)): $x_4^3 \equiv 7(\text{mod } 10)$ whence follows that $x_4 = 3$ as can be verified from (17). Thus the last digit of the cube root can at once be determined.

Again

$$x_1 = [\sqrt[3]{927}] = 9$$

So there remain only $x_2$ and $x_3$

$$X \equiv x_3.10 + x_4(\text{mod } 10^2), \quad A \equiv 77(\text{mod } 10^2)$$

therefore

$$(x_3.10 + x_4)^3 \equiv 77(\text{mod } 10^2)$$
$$(x_3.10 + 3)^3 \equiv 77(\text{mod } 10^2)$$
$$x_3^3.10^3 + 3.x_3^2.10^2.3 + 3(x_3.10)3^2 + 3^3 \equiv 77(\text{mod } 10^2)$$

Since terms $\equiv 0(\text{mod } 10^2)$ can be neglected from this congruence

$$3^3 + 3.(x_3.10)3^2 \equiv 77(\text{mod } 10^2)$$
$$270x_3 \equiv 50(\text{mod } 10^2)$$
$$70x_3 \equiv 50(\text{mod } 10^2)$$
$$7x_3 \equiv 5(\text{mod } 10)$$

Since (7,10) = 1 this congruence has a unique (mod 10) solution namely

$$x_3 \equiv 5(\text{mod } 10)$$

But $x_3$ being a digit of $X$, one has $0 \le x_3 \le 9$. Consequently $x_3$ must be *equal* to 5.

Now $b_1 = a_1 - x_1^3$, $a_1 = 927$, $x_1 = 9$, thus $b_1 = 198$ and we know that

$$x_2 = \left[\frac{1987}{3x_1^2}\right] = \left[\frac{1987}{243}\right] = 7$$

Hence $(x_1, x_2, x_3, x_4) = (9, 7, 5, 3)$

However we must point out that this procedure *assumes* that $A$ is a perfect cube. Accordingly it becomes necessary to verify that $A = (9753)^3$.

*Remarks:* $1^0$ An analysis of the proof of Theorem 2 reveals that at each stage **I⁰, II⁰, III⁰**, in (12) we are removing the largest cube from the corresponding number, i.e.

$$r_1 = [\sqrt[3]{a_1}]$$
$$r_2 = [\sqrt[3]{A_1}], \quad (A_1 = a_1.10^3 + a_2.10^2 + a_3.10 + a_4)$$
$$r_3 = [\sqrt[3]{A_1.10^3 + A_2}], \quad (A_2 = a_5.10^2 + a_6.10 + a_7)$$
$$r_4 = [\sqrt[3]{A_1.10^6 + A_2.10^3 + A_3}], \quad (A_3 = a_8.10^2 + a_9.10 + a_{16})$$

Thus, in Example 3, [See p. 96],

$$9 = [\sqrt[3]{792}]$$
$$92 = [\sqrt[3]{792994}]$$
$$925 = [\sqrt[3]{792994249}]$$

(where, as usual, [...] stands for integral part of a number).

$2^0$ Let us use congruences for the number we worked with in Example 1: $A = 953\ 125$. If $X = x_1.10^2 + x_2.10 + x_3$ we have

$$x_1 = [\sqrt[3]{1}] = 1, x_3^3 \equiv 5(\text{mod}\, 10), \text{ so that } x_3 = 5$$

Next we have

$$[x_2.10 + x_3]^3 \equiv 25(\text{mod}\, 10^2)$$
$$[x_2.10 + 5]^3 \equiv 25(\text{mod}\, 10^2)$$
$$5^3 + 3.5^2.(x_2.10) \equiv 25(\text{mod}\, 10^2),$$

(where terms in the expansion of the cube are neglected whenever these are $\equiv 0$ (mod $10^2$). Thus

$$125 + 750x_2 \equiv 25(\text{mod}\, 10^2)$$
$$50x_2 \equiv 0(\text{mod}\, 10^2)$$
$$x_2 \equiv 0(\text{mod}\, 2)$$

But $x_2$ being a digit must satisfy $0 \leq x_2 \leq 9$, thus

$$x_2 = 0, 2, 4, 6, 8$$

Since $12^3 < 1953 < 13^3, [\sqrt[3]{1953}] = 12$, so $x_2 \neq 0, 4, 6, 8$, thus $x_2 = 2$. Congruences have not been convenient in this example.

**Example 5.** A 13-digit perfect cube begins with 5 and ends with 0541. Find it and its cube root.

We are given

$$A = \langle a_1 \quad a_2a_3a_4 \quad a_5a_6a_7 \quad a_8a_9a_{10} \quad a_{11}a_{12}a_{13} \rangle$$
$$a_1 = 1$$
$$a_{10} = 0$$
$$a_{11} = 5$$
$$a_{12} = 4$$
$$a_{13} = 1$$

For the cube root $X$ of $A$, let $X = \langle x_1, x_2, x_3, x_4, x_5 \rangle$

$$X = x_1.10^4 + x_2.10^3 + x_3.10^2 + x_4.10 + x_5$$

Then

$$\left.\begin{array}{r} X \equiv x_5(\text{mod}\, 10) \\ A \equiv 1(\text{mod}\, 10) \end{array}\right\} \Rightarrow x_5^3 \equiv 1(\text{mod}\, 10), \text{ thus } \underline{x_5 = 1}$$

Similarly,

$$(x_4.10 + x_5)^3 \equiv 41(\text{mod}\, 10^2)$$

$$(x_4.10+1)^3 \equiv 41(\text{mod}\,10^2)$$
$$1+3(x_4.10) \equiv 41(\text{mod}\,10^2)$$

$$30x_4 = 40(\text{mod}\,10^2), \text{ so that } 3x_4 \equiv 4(\text{mod}\,10)$$

This gives $\underline{x_4 = 8}$

Again

$$(x_3.10^2 + x_4.10 + x_5)^3 \equiv 541(\text{mod}\,10^3)$$
$$(x_3.10^2 + 8.10 + 1)^3 \equiv 541(\text{mod}\,10^3)$$
$$(x_3.10^2 + 81)^3 \equiv 541(\text{mod}\,10^3)$$
$$81^3 + 3.81^2(x_3 10^2) \equiv 541(\text{mod}\,10^3)$$

Since

$$81 \equiv 1(\text{mod}\,10), \quad 81^2 \equiv 1(\text{mod}\,10), \quad 10^2.81^2 \equiv 10^2(\text{mod}\,10^3),$$
$$3.10^2.81^2 \equiv 3.10^3(\text{mod}\,10^3)$$

and

$$81^2 \equiv 561(\text{mod}\,10^3), 81^3 \equiv 441(\text{mod}\,10^3)$$

we obtain that

$$441 + 300x_3 = 541(\text{mod}\,103)$$
$$300x_3 \equiv 100(\text{mod}\,10^3)$$
$$3x_3 \equiv 1(\text{mod}\,10), \quad \text{i.e } \underline{x_3 = 7}$$

Next we have $A \equiv 0541(\text{mod}\,10^4)$, $X \equiv (x_2.10^3 + 781)\ (\text{mod}\,10^4)$

Hence

$$(x_2 10^3 + 781)^3 \equiv 0541(\text{mod}\,10^4)$$
$$(781)^3 + 3.(781)^2.x_2.10^3 \equiv 541(\text{mod}\,10^4)$$

$$9541 + 3.10^3 x_2 \equiv 541(\text{mod}\,10^4)$$

$$9.10^3 + 3.10^3 x_2 \equiv 0(\text{mod}\,10^4) \qquad 781^2 \equiv 1(\text{mod}\,10)$$

$$9 + 3x_2 \equiv 0(\text{mod}\,10) \qquad (781)^2.10^3 \equiv 10^3(\text{mod}\,10^4)$$

$$\therefore x_2 = 7 \qquad 3.10^3.(781)^2 \equiv 3.10^3(\text{mod}\,10^4)$$

Consequently

$$\sqrt{A} = X = 17781$$

$$\begin{aligned} 781^3 &= (7.10^2 + 81)^3 \\ &\equiv 3.(7.10^2)81^2 + 81^3 (\text{mod}\, 10^4) \\ &\equiv 21.10^2.(6561) + 1441 (\text{mod}\, 10^4) \\ &\equiv 9541 (\text{mod}\, 10^4) \end{aligned}$$

In order to find A we write

$$A = (177.10^2 + 81)^3$$

We compute $177^2$, $177^3$, and $81^2, 81^3$ and find

$$\underline{A = 5\ 621\ 731\ 190\ 541}$$

[See ≪VEDIC MATHEMATICS≫ P.325 for a different procedure]

**Example 6.** Find the cube root of $A = 780\ 212\ 513\ 816$

[Compare Example 3, p. 96]

| $k$ | 1 | 2 | 3 | 4 | 5 | 6 | 7 | 8 | 9 | 10 |
|---|---|---|---|---|---|---|---|---|---|---|
| $a_k$ | 780 | 2 | 1 | 2 | 5 | 1 | 3 | 8 | 1 | 6 |
| $b_k$ | 51 | 26 | 153 | 1524 | 15245 | 152451 | 1524513 | 9938 | 21 | 0 |
| $x_k$ | 9 | 2 | 0 | 6 | | | | | | |
| $r_k$ | 9 | 92 | 920 | 9206 | | | | | | |

$$\sqrt[3]{A} = 9206$$

In the notation of Theorem 2 (P.94) we have

$$A_1 = 780212, \quad A_2 = 513, \quad A_3 = 816$$
$$3r_2^2 = 3 \times 92^2 = 3 \times 464 = 25392$$

Hence for the group **II⁰** of equations on p.94, wherein we have to divide $10b_4 + a_5$ by $3r_2^2$, we have $10b_4 + a^5 < 3r_2^2$, i.e. we have to take the quotient $x_3 = 0$ which makes

$$10b_4 + a_5 = b_5, 10b_5 + a_6 = b_6, 10b_6 + a_7 = b_7$$

CHAPTER III

# THE BRAHMAGUPTA-BHASKARA EQUATION

## § 1. LEMMAS OF BRAHMAGUPTA

We consider quadratic indeterminate equations of the form

$$\boxed{Dx^2 \pm c = y^2} \qquad (D > 0) \tag{1}$$

Herein $D$ and $c$ are given *integers*, $D \neq$ *a* square integer. The problem is to find *positive integer* solutions $(x, y)$ of the equation (1). Bhaskara II (1150 A.D.) deals with this problem extensively in his treatise Beeja Ganitha [See Chapter VI: Varga Prakriti and Chapter VII - Chakravala].

There are some important lemmas of Brahmagupta which are basic for the following discussion.

**LEMMA 1 [BRAHMAGUPTA]**

If $(x, y) = (\alpha, \beta)$ is a solution of the equation

$$Dx^2 + k = y^2, \tag{2}$$

and if $(x, y) = (\alpha', \beta')$ is a solution of the equation

$$Dx^2 + k' = y^2 \tag{3}$$

then

$$(x, y) = \left| \begin{array}{l} (\alpha\beta' + \alpha'\beta, \beta\beta' + D\alpha\alpha') \\ \\ (\alpha\beta' - \alpha'\beta, \beta\beta' - D\alpha\alpha') \end{array} \right.$$

are solutions of the equation

$$Dx^2 + kk' = y^2$$

**Proof:** We have

$$D\alpha^2 + k = \beta^2, \quad D\alpha'^2 + k' = \beta'^2$$

Taking

$$(x, y) = (\alpha\beta' + \alpha'\beta, \beta\beta' + D\alpha\alpha')$$

we see that

$$\begin{aligned}y^2 - Dx^2 &= (\beta\beta' + D\alpha\alpha')^2 - D(\alpha\beta' + \alpha'\beta)^2 \\ &= (\beta^2\beta'^2 + D^2\alpha^2\alpha'^2 + 2\beta\beta'.D\alpha\alpha')- \\ &\qquad - D(\alpha^2\beta'^2 + \alpha'^2\beta^2 + 2\alpha\beta'\alpha'\beta) \\ &= \beta'^2(\beta^2 - D\alpha^2) - D\alpha'^2(\beta^2 - D\alpha^2) \\ &= (\beta^2 - D\alpha^2).(\beta'^2 - D\alpha'^2) = kk'\end{aligned}$$

Similarly it can be verified that

$$(x, y) = (\alpha\beta' - \alpha'\beta, \beta\beta' - D\alpha\alpha')$$

satisfies the equation $Dx^2 + kk' = y^2$.

**LEMMA 2 [BRAHMAGUPTA]**

If $(\alpha, \beta)$ is a solution of the equation

$$Dx^2 + k = y^2$$

then $(x, y) = (2\alpha\beta, \beta^2 + \Delta\alpha^2)$ is a solution of the equation

$$Dx^2 + k = y^2$$

**Proof:** This follows from Lemma 1 when we take

$$k' = k \text{ and } (\alpha'\beta') = (\alpha, \beta)$$

It can also be directly verified as follows:

$$\begin{aligned}y^2 - Dx^2 &= (\beta^2 + D\alpha^2)^2 - D.(2\alpha\beta)^2 \\ &= (\beta^2 + D\alpha^2)^2 - 4.\beta^2.D\alpha^2 \\ &= (\beta^2 - D\alpha^2)^2 = k^2,\end{aligned}$$

since $D\alpha^2 + k = \beta^2$.

**LEMMA 3 [BRAHMAGUPTA]**

If $(x, y) = (\alpha, \beta)$ is a solution of the equation

$$Dx^2 + k = y^2$$

such that $k|\alpha$ and $k|\beta$ then $\left(\dfrac{\alpha}{k}, \dfrac{\beta}{k}\right)$ is a solution of the equation

$$Dx^2 + 1 = y^2$$

**Proof:** We have that $\alpha' = \dfrac{\alpha}{k}$ and $\beta' = \dfrac{\beta}{k}$ are integers, and

$$\beta'^2 - D\alpha'^2 = \frac{\beta^2}{k^2} - D \cdot \frac{\alpha^2}{k^2} = \frac{1}{k^2}(\beta^2 - D\alpha^2) = 1$$

since $Dx^2 + k = \beta^2$.

*Remark* If we take $k = 1$ in Lemma 2 we see that

(6) If $(\alpha, \beta)$ is a solution of $Dx^2 - 1 = y^2$, then $(2\alpha\beta, \beta^2 + D\alpha^2)$ is a solution of $Dx^2 + 1 = y^2$.

We shall see later that the equation

$$Dx^2 + 1 = y^2$$

is *always* possible. However the equation

$$Dx^2 - 1 = y^2$$

may have no integer solutions.

**Example**

The equation $3x^2 - 1 = y^2$ has no solutions in integers.

We have, always, for any integer $x$

$$x^2 \equiv \begin{cases} 0(mod 4) \text{ for } x \text{ even} \\ 1(mod 4) \text{ for } x \text{ odd} \end{cases}$$

so that

$$3x^2 \equiv \begin{cases} 0(mod 4) \text{ for } x \text{ even} \\ 3(mod 4) \text{ for } x \text{ odd} \end{cases}$$

and

$$y^2 + 1 \equiv \begin{cases} 1(mod 4) \text{ for } y \text{ even} \\ 2(mod 4) \text{ for } y \text{ odd} \end{cases}$$

Consequently, for all integers $x$ and $y$

$$3x^2 \neq y^2 + 1$$

since they must lie in different residue classes (mod 4).

The same argument holds even when we have, instead of 3, any number $D \equiv 3(mod 4)$

Consequently:

(7) The equation $Dx^2 - 1 = y^2$ has *no* integer solutions whenever $D \equiv 3(mod 4)$

*Remark*

Brahmagupta's lemmas have been described by Bhaskara II (1150), Narayana (1350), Jnana Raja (1503) and Kamalakara (1658).

The above lemmas are referred to by their technical designation: ≪BHĀVANĀ≫.

Euler (1764) and Lagrange (1768) rediscovered Brahmagupta's lemmas, and recognised their fundamental nature.

## § 2. EXAMPLES

**Example 1.** (Bhaskara II)

$1^0$ What is that square which multiplied by 8 becomes, together with unity, a square?

$2^0$ What square, multiplied by 11 and increased by unity, becomes a square?

We have to solve

$$1^0 \quad 8x^2 + 1 = y^2 \text{ and } 2^0 \quad 11x^2 + 1 = y^2$$

$1^0$ $\underline{8x^2 + 1 = y^2, (D = 8)}$

By inspection $(x_0, y_0) = (1, 3)$ is a solution of $1^0$:

$$8x_0^2 + 1 = 8 + 1 = 9 = y_0^2$$

By Lemma 2

$$(x_1, y_1) = (2x_0y_0, Dx_0^2 + y_0^2) = (6, 17)$$

is again a solution of $1^0$. Then by Lemma 1 we can obtain a new solution by the composition (Bhāvanā) of the two solutions $(x_0, y_0) = (1, 3)$ and $(x_1, y_1) = (6, 17)$, a new solution $(x_2, y_2)$:

$$(x_2, y_2) = (x_0y_1 + x_1y_0, Dx_0x_1 + y_0y_1) = (35, 99)$$

In this manner we have an infinte number of solutions

$2^0$ $\underline{11x^2 + 1 = y^2, \quad (D = 11)}$

Taking $x_0 = 1$ we observe that $11x_0^2 = 11 = 3^2 + 2$ which means that:

$(x_0, y_0) = (1, 3)$ is a solution of the *auxiliary* equation

$$11x^2 - 2 = y^2$$

By Lemma 2, if we put

$$(x_1, y_1) = (2x_0y_0, Dx_0^2 + y_0^2)$$

then $(x_1, y_1) = (6,20)$ satisfies the relation

$$11x_1^2 + 4 = y_1^2 \tag{8}$$

Taking $(x_2, y_2) = \left(\frac{x_1}{2}, \frac{y_1}{2}\right) = (3, 10)$ we have from (8):

$$11 \cdot (2x_2)^2 + 4 = (2y_2)^2$$

$$11x_2^2 + 1 = y_2^2$$

Thus $(x_2, y_2)$ = (3,10) is a solution of the equation $2^0$. We obtain, by lemma 2, a new solution:

$$(x_3, y_3) = (2x_2y_2, Dx_2^2 + y_2^2) = (60, 199)$$

Using two solutions $(x_0, y_0) = (1, 3)$ and $(x_3, y_3) = (60, 199)$. We can obtain, by repeated application of the lemmas $1^0$ and $2^0$ an infinte number of solutions.

**Example 2.** [*Rational* solutions of the equation $11x^2 + 1 = y^2$]. The real interest lies in the search for positive integer solutions. If $m$ is any *rational* number, then

$$(x, y) = \left(\frac{2m}{m^2 - D}, \frac{m^2 + D}{m^2 - D}\right)$$

will always be a solution of the equation

$$Dx^2 + 1 = y^2$$

In fact,

$$\begin{aligned} y^2 - Dx^2 &= \frac{1}{(m^2 - D)^2}\left\{(m^2 + D)^2 - D \cdot (2m)^2\right\} \\ &= \frac{(m^2 - D)^2}{(m^2 - D)^2} = 1 \end{aligned}$$

We can find the rational solutions of the equation $11x^2 + 1 = y^2$ as follows. Taking $x_0 = 1$ we have

$$11x_0^2 + 5 = 11 + 5 = 4^2$$

i.e., $(x_0, y_0) = (1, 4)$ is a solution of the auxiliary equation

$$11x^2 + 5 = y^2$$

By lemma 2, $(x_1, y_1) = (2x_0y_0, Dx_0^2 + y_0^2) = (8, 27)$ is a solution of the equation

$$11x^2 + 25 = y^2$$

Put $(x_2, y_2) = \left(\frac{x_1}{5}, \frac{y_1}{5}\right)$. Then

$$11x_1^2 + 25 = y_1^2, 11\left(\frac{x_1}{5}\right)^2 + 1 = \left(\frac{y_1}{5}\right)^2,$$

i.e., $(x_2, y_2) = \left(\frac{8}{5}, \frac{27}{5}\right)$ is a *rational* solution of the equation $11x^2 + 1 = y^2$. Since (3,10) is another solution of this equation we can get, by composition of the two solution (3,10) and $\left(\frac{8}{5}, \frac{27}{9}\right)$ still other solutions $(x_3, y_3)$ and $(x_3', y_3')$ by lemma 1.

$$\begin{cases} x_3 = 3\left(\frac{27}{5}\right) + 10\left(\frac{8}{5}\right) \\ y_3 = 10\left(\frac{27}{5}\right) + 11\cdot 3\left(\frac{8}{5}\right) \end{cases} \quad \text{and} \quad \begin{array}{l} x_3' = 3\left(\frac{27}{5}\right) - 10\left(\frac{8}{5}\right) \\ y_3' = 10\left(\frac{27}{5}\right) - 11\cdot 3\left(\frac{8}{5}\right) \end{array}$$

thus

$$(x_3, y_3) = \left(\frac{161}{5}, \frac{534}{5}\right) \text{ and } (x_3', y_3') = \left(\frac{1}{5}, \frac{6}{5}\right)$$

are *rational* solutions of the equation $11x^2 + 1 = y^2$.

We have seen that Brahmagupta's method for the general solution in positive integers of the equation.

$$Dx^2 + 1 = y^2, \quad D \neq \text{ a square integer}$$

depends on finding a solution of an *auxiliary* equation of the form

$$Dx_0^2 + k = y_0^2, \quad (k = \pm 1, \pm 2, \pm 4)$$
$$(x_0, y_0 \text{ positive integers })$$

Then by repeated application of Brahmagupta's lemmas any number of new solutions can be obtained.

## § 3. CHAKRAVALA METHOD OF BHASKARA

Brahmagupta had no method of finding an auxiliary equation of the required form. It was Bhaskara (II) who, by his elegant and powerful method of ≪ Chakravala≫ (or ≪ cyclic method ≫) showed how to construct an auxiliary equation with its two integer solutions.

Before we prove the theorem of Bhaskara we shall illustrate the basic ideas by a few examples. First we prove:

**LEMMA** (Bhaskara)

If $F(x, y) = Dx^2 - y^2$ and $h_0 = F(x_0, y_0)$, and

(1) $$\boxed{(x_1, y_1, h_1) = \left(\frac{m_0 x_0 + y_0}{h_0}, \frac{Dx_0 + m_0 y_0}{h_0}, \frac{D - m_0^2}{h_0}\right)}$$

then one has the identity.

(2) $$\boxed{Dx_1^2 - y_1^2 = -h_1}$$

i.e.,

(3) $$\boxed{F(x_1, y_1) = -h_1}$$

We shall refer to (2) as *Bhaskara's identity*

**Proof:** We have

$$F(x_1, y_1) = Dx_1^2 - y_1^2 = D\left(\frac{m_0x_0 + y_0}{h_0}\right)^2 - \left(\frac{Dx_0 + m_0y_0}{h_0}\right)^2$$

$$= \frac{1}{h_0^2}\Big[D(m_0^2x_0^2 + 2m_0x_0.y_0 + y_0^2)$$

$$- (D^2x_0^2 + 2Dx_0.m_0y_0 + m_0^2y_0^2)\Big]$$

$$= \frac{1}{h_0^2}\Big[Dx_0^2(m_0^2 - D) - y_0^2(m_0^2 - D)\Big]$$

$$= \frac{1}{h_0^2}(Dx_0^2 - y_0^2)(m_0^2 - D)$$

$$= \frac{1}{h_0^2} \cdot h_0(m_0^2 - D)$$

$$= -\frac{D - m_0^2}{h_0} = -h_1$$

We shall make the assumption that $x_0, y_0$ are coprime: $(x_0, y_0) = 1$. Since $h_0 = F(x_0, y_0) = Dx_0^2 - y_0^2$, it follows that $x_0, y_0$ and $h_0$ are mutually coprime. If we determine the positive integer $m_0$ such that

(4) $$\boxed{m_0x_0 + y_0 \equiv 0(\text{mod}\, h_0)}$$

then $x_1 = \dfrac{m_0x_0 + y_0}{h_0}$ is an integer. We claim that $y_1$ and $h_1$ in (1) are also then integers. In fact, since $h_0 = Dx_0^2 - y_0^2$ we have

$$Dx_0^2 \equiv y_0^2(\text{mod}\, h_0)$$

By (4)

$$m_0x_0 \equiv -y_0(\text{mod}\, h_0)$$

Multiplying the latter congruence by $y_0$ and adding to the previous one, we get,

$$Dx_0^2 + m_0x_0y_0 \equiv 0(\text{mod}\, h_0)$$

$$x_0(Dx_0 + m_0y_0) \equiv 0(\text{mod}\, h_0)$$

Since $(x_0, h_0) = 1$, this implies

(5) $$\boxed{Dx_0 + m_0y_0 \equiv 0(\text{mod}\, h_0)}$$

i.e., $y_1 = \dfrac{Dx_0 + m_0 y_0}{h_0}$ is an integer.

Again, from (4), $y_0 \equiv -m_0 x_0 (\operatorname{mod} h_0)$; substituting this in (5) we get

$$Dx_0 - m_0^2 \equiv 0 (\operatorname{mod} h_0), \quad (D - m_0^2)x_0 \equiv (\operatorname{mod} h_0).$$

Since $(x_0, h_0) = 1$, this implies that

(6) $$\boxed{D - m_0^2 \equiv 0 (\operatorname{mod} h_0)}$$

i.e. $h_1 = \dfrac{D - m_0^2}{h_0}$ is an integer.

Next we observe that

$$x_0 y_1 - x_1 y_0 = x_0 \left( \frac{Dx_0 + m_0 y_0}{h_0} \right) - \left( \frac{m_0 x_0 + y_0}{h_0} \right) y_0$$
$$= \frac{1}{h_0}(Dx_0^2 - y_0^2) = 1$$

i.e.,

(7) $$\boxed{x_0 y_1 - x_1 y_0 = 1}$$

Consequently, $x_1, y_1, h_1$ must be mutually coprime.

### BHASKARA's method of solving the equation $Dx^2+1=y^2$ $(D > 0,\ \ D \neq a$ square integer)

As always, $F(x, y) = Dx^2 - y^2$. Put

(8) $$\boxed{x_0 = 1, \ \ y_0 = [\sqrt{D}], \ \ h_0 = F(x_0, y_0)} \qquad (h_0 > 0)$$

If we put, as in (1),

$$(x_1, y_1, h_1) = \left( \frac{m_0 x_0 + y_0}{h_0}, \frac{Dx_0 + m_0 y_0}{h_0}, \frac{D - m_0^2}{h_0} \right)$$

then, by Bhaskara's identity

(9) $$\boxed{F(x_1, y_1) = -h_1} \qquad (h_1 > 0)$$

So far, the *multiplier* $m_0$ has been subject to the sole restriction that $m_0$ must satisfy the congruence

(10) $$\boxed{m_0 x_0 + y_0 \equiv 0 (\operatorname{mod} h_0)}$$

We now stipulate that the multiplier $m_0$ must satisfy the inequalities:

(11) $$\boxed{m_0 < \sqrt{D} < m_0 + h_0}$$

We can now proceed with $(x_1, y_1, h_1)$ as we did with $(x_0, y_0, h_0)$. In this manner, we can define, inductively, the four sequences of positive integers

$$(m_\lambda, x_\lambda, y_\lambda, h_\lambda) \quad (\lambda = 0, 1, 2, \cdots)$$

with the properties:

$$1^0 \quad h_\lambda = (-1)^\lambda F(x_\lambda, y_\lambda)$$

$$(12) \quad 2^0 \quad (x_{\lambda+1}, y_{\lambda+1}, h_{\lambda+1}) = \left(\frac{m_\lambda x_\lambda + y\lambda}{h_\lambda}, \frac{Dx_\lambda + m_\lambda y\lambda}{h_\lambda}, \frac{D - m_\lambda^2}{h_\lambda}\right)$$

$$3^0 \quad m_\lambda x_\lambda - y_\lambda \equiv 0 (\operatorname{mod} h_\lambda)$$

$$4^0 \quad m_\lambda < \sqrt{D} < m_\lambda + h_\lambda$$

Then, the claim of Bhaskara's theorem is:

> There exists $N$ such that $h_{N-1} = 1$,
> i.e., $F(x_{N-1}, y_{N-1}) = (-1)^N$

i.e., $(x_{N-1}, y_{N-1})$ is a solution of the equation.

$$Dx^2 - y^2 = (-1)^N$$

If $N$ is odd, this means that we can solve

$$Dx^2 + 1 = y^2$$

If $N$ is even, we can solve

$$Dx^2 - 1 = y^2$$

but then, (See (6)p.105)the equation $Dx^2 + 1 = y^2$ can also be solved.

**Example 1.** [BHASKARA II]

What is that number whose square, multiplied by 67 and then added to unity, yields a perfect square?

We have to solve

$$Dx^2 + 1 = y^2, \quad \underline{D = 67}$$

Since $8^2 < D < 9^2, 8 < \sqrt{D} < 9$, we have $[\sqrt{D}] = 8$.
Take $(x_0, y_0) = (1, 8)$; then

$$h_0 = Dx_0^2 - y_0^2 = 67 - 64 = 3$$

thus

$$(13) \qquad (x_0, y_0, h_0) = (1, 8, 3)$$

In order to find $(x_1, y_1, h_1)$ we solve the congruence

$$mx_0 + y_0 \equiv 0 (\operatorname{mod} h_0)$$

$$m + 8 \equiv 0 (\operatorname{mod} 3)$$

$$m \equiv 1 (\operatorname{mod} 3)$$

In the arithmetic progression

$$m = 1, 4, 7, 10, ...$$

we find

$$7 < \sqrt{D} < 10$$

Hence $\underline{m_0 = 7}$. Then

$$x_1 = \frac{m_0 x_0 + y_0}{h_0} = \frac{7+8}{3} = 5$$

$$y_1 = \frac{Dx_0 + my_0}{h_0} = \frac{67+56}{3} = 41$$

$$h_1 = \frac{D - m_0^2}{h_0} = \frac{67-49}{3} = 6$$

thus

(14) $$(x_1, y_1, h_1) = (5, 41, 6)$$

Now, solving the congruence

$$mx_1 + y_1 \equiv 0 (\operatorname{mod} h_1)$$

$$5m + 41 \equiv 0 (\operatorname{mod} 6)$$

$$5m \equiv 1 (\operatorname{mod} 6)$$

$$m \equiv 5 (\operatorname{mod} 6)$$

In the arithmetic progression

$$m = 5, 11, 16, ...$$

We have

$$5 < \sqrt{D} < 11$$

Hence $\underline{m_1 = 5}$. Then

$$x_2 = \frac{m_1 x_1 + y_1}{h_1} = \frac{25+41}{6} = 11$$

$$y_2 = \frac{Dx_1 + m_1 y_1}{h_1} = \frac{335+205}{6} = 90$$

$$h_2 = \frac{D - m_1^2}{h_1} = \frac{67-25}{6} = 7$$

(15) $$(x_2, y_2, h_2) = (11, 90, 7)$$

Now, solving the congruence

$$\begin{aligned} mx_2 + y_2 &\equiv 0 (\operatorname{mod} h_2) \\ 41m + 90 &\equiv 0 (\operatorname{mod} 7) \\ 4m &\equiv 1 (\operatorname{mod} 7) \\ m &\equiv 2 (\operatorname{mod} 7) \end{aligned}$$

In the arithmetic progression

$$m = 2, 9, 16, ...$$

we have $2 < \sqrt{D} < 9$, thus $m_2 = 2$. Then

$$\begin{aligned} x_3 &= \frac{m_2 x_2 + y_2}{h_2} = \frac{22 + 90}{7} = 16 \\ y_3 &= \frac{D x_2 + m_2 y_2}{h_2} = \frac{737 + 180}{7} = 131 \\ h_3 &= \frac{D - m_2^2}{h_2} = \frac{67 - 4}{7} = 9 \end{aligned}$$

(16) $$(x_3, y_3, h_3) = (16, 131, 9)$$

Now, solving the congruence

$$\begin{aligned} mx_3 + y_3 &\equiv 0 (\operatorname{mod} h_3) \\ 16m + 131 &\equiv 0 (\operatorname{mod} 9) \\ 7m + 5 &\equiv 0 (\operatorname{mod} 9) \\ m &\equiv 7 (\operatorname{mod} 9) \end{aligned}$$

In the arithmetic progression

$$m = 7, 16, 25, ...$$

We have $7 < \sqrt{D} < 16$, thus $\underline{m_3 = 7}$. Then

$$\begin{aligned} x_4 &= \frac{m_3 x_3 + y_3}{h_3} = \frac{112 + 131}{9} = 27 \\ \text{(17)}\quad y_4 &= \frac{D x_3 + m_3 y_3}{h_3} = \frac{1072 + 917}{9} = 221 \\ h_4 &= \frac{D - m_3^2}{h_3} = \frac{67 - 49}{9} = 2 \end{aligned}$$

Now, as in the proof of Bhaskara's lemma [See p. 108] we find

$$h_\lambda = (-1)^\lambda (D x_\lambda^2 - y_\lambda^2) = (-1)^\lambda F(x_\lambda \cdot y_\lambda)$$

By (17), we have $h_4 = 2$, so

(18) $\boxed{Dx_4^2 - y_4^2 = 2}$ with $(x_4, y_4) = (27, 221)$

As soon as we obtain $h_\lambda = \pm 2, \pm 4$ we can use Brahmagupta's lemmas on composition of solutions to solve $Dx^2 + 1 = y^2$. In fact, it follows from (18) that $(x_4', y_4') = (2x_4', y_4', Dx_4^2 + y_4^2) = (11934, 97684)$ satisfies the relation

$$Dx_4'^2 + 4 = y_4'^2$$

Consequently,

$$\left(\frac{x_4'}{2}, \frac{y_4'}{2}\right) = (5967, 48842)$$

satisfies the equation $Dx^2 + 1 = y^2$.

(19) $\boxed{\begin{array}{c} 67x^2 + 1 = y^2 \\ \text{(x,y) = (5967, 48842)} \end{array}}$

**Example 2.** [BHASKARA II]

> What is that number whose square multiplied by 61 and then added to 1, yields a perfect square?

We have to solve

$$Dx^2 + 1 = y^2, \quad \underline{D = 61}$$

Since $7 < \sqrt{61} < 8$, we take $(x_0, y_0) = (1, 7)$. Then $h_0 = Dx_0^2 - y_0^2 = 61 - 49 = 12$. Solving the congruence

$$\begin{aligned} mx_0 + y_0 &\equiv 0 (\text{mod}\, h_0) \\ m + 7 &\equiv 0 (\text{mod}\, 12) \\ m &\equiv 5 (\text{mod}\, 12) \end{aligned}$$

In the sequence $m = 5, 17, \cdots$, we must take $m_0 = 5$.

Hence

$$(m_0, x_0, y_0, h_0) = (5, 1, 7, 12)$$

We then find

$$\left.\begin{aligned} x_1 &= \frac{m_0 x_0 + y_0}{h_0} = 1 \\ y_1 &= \frac{Dx_0 + m_0 y_0}{h_0} = 8 \\ h_1 &= \frac{D - m_0^2}{h_0} = 3 \end{aligned}\right| \quad (x_1, y_1, h_1) = (1, 8, 3)$$

Next we solve the congruence

$$mx_1 + y_1 \equiv 0 (\text{mod}\, h_1)$$
$$m + 8 \equiv 0 (\text{mod}\, 3)$$
$$m \equiv 1 (\text{mod}\, 3)$$

In the sequence $m = 1, 4, 7, 10, \cdots$ we must take $m_1 = 7$. We then find: $(m_1, x_1, y_1, h_1) = (7, 1, 8, 3)$ and

$$x_2 = \frac{m_1 x_1 + y_1}{h_1} = 5$$
$$y_2 = \frac{Dx_1 + m_1 y_1}{h_1} = 39 \qquad (x_2, y_2, h_2) = (5, 39, 4)$$
$$h_2 = \frac{D - m_1^2}{h_1} = 4$$

Since $h_\lambda = (-1)^\lambda F(x_\lambda, y_\lambda)$, and $h_2 = 4$ we have

$$Dx_2^2 - y_2^2 = 4$$

or

(20) $\boxed{Dx_2^2 - 4 = y_2^2}$ with $(x_2, y_2) = (5, 39)$

It follows from (20) that $(\alpha, \beta) = \left(\frac{x_2}{2}, \frac{y_2}{2}\right) = \left(\frac{5}{2}, \frac{39}{2}\right)$ satisfies the equation

(21) $$D\alpha^2 - 1 = \beta^2, (\alpha, \beta) = \left(\frac{5}{2}, \frac{39}{2}\right)$$

Composing the solution $(\alpha, \beta)$ with itself we see that

$$(\alpha_1, \beta_1) = (2\alpha\beta, D\alpha^2 + \beta^2) = \left(\frac{195}{2}, \frac{1523}{2}\right)$$

must satisfy the equation

(22) $$D\alpha_1^2 + 1 = \beta_1^2, (\alpha_1, \beta_1) = \left(\frac{195}{2}, \frac{1523}{2}\right)$$

Applying Brahmagupta's lemma to (21) and (22) it follows that

$$(\xi, \eta) = (\alpha\beta_1 + \alpha_1\beta_1, D\alpha\alpha_1 + \beta\beta_1)$$
$$= (3805, 29718)$$

must satisfy the relation

(22′) $\boxed{D\xi^2 - 1 = \eta^2}$ $(\xi, \eta) = (3805, 29718)$

Finally we compose $(\xi, \eta)$ with itself to conclude that

$$\begin{aligned}(\xi_0, \eta_0) &= (2\xi\eta, D\xi^2 + \eta^2)\\ &= (226153910, 1766319049)\end{aligned}$$

satisfies the equation $Dx^2 + 1 = y^2$:

$$\boxed{\begin{array}{l}61x^2 + 1 = y^2\\ (x, y) = (226153910, 1766319049)\end{array}}$$

**Example 3.** [Narayana (1356)]

The equation

$$\boxed{Dx^2 + 1 = y^2 \text{ with } D = 103}$$

Since $10 < \sqrt{D} < 11$ we take $(x_0, y_0) = (1, 10)$ so that

$$h_0 = Dx_0^2 - y_0^2 = 3$$

*Step 1:*

Solving the congruence $mx_0 + y_0 \equiv 0(\text{mod}\, h_0)$ we have:
$m + 10 \equiv 0(\text{mod}\, 3), \quad m \equiv 2(\text{mod}\, 3)$,
$m = 2, 5, 8, 11, \cdots$; we must take $\underline{m_0 = 8}$.

Thus

$$(m_0, x_0, y_0, h_0) = (8, 1, 10, 3)$$

Hence

$$\begin{aligned}x_1 &= \frac{m_0x_0 + y_0}{h_0} = 6,\\ y_1 &= \frac{Dx_0 + m_0y_0}{h_0} = 61,\\ h_1 &= \frac{D - m_0^2}{h_0} = 13\end{aligned}$$

i.e.,

$$(x_1, y_1, h_1) = (6, 61, 13)$$

*Step 2:*

$$\begin{aligned}mx_1 + y_1 &\equiv 0(\text{mod}\, h_1)\\ 6m + 61 &\equiv 0(\text{mod}\, 13)\\ 6m + 9 &\equiv 0(\text{mod}\, 13)\\ m &\equiv 5(\text{mod}\, 13), \quad m = 5, 18, 31, \cdots\end{aligned}$$

We must take $\underline{m_1 = 5} : (m_1, x_1, y_1, h_1) = (5, 6, 61, 13)$

Hence

$$x_2 = \frac{m_1 x_1 + y_1}{h_1} = 7,$$

$$y_2 = \frac{D x_1 + m_1 y_1}{h_1} = 71,$$

$$h_2 = \frac{D - m_1^2}{h_1} = 6$$

$$(x_2, y_2, h_2) = (7, 71, 6)$$

*Step 3:*

$$mx_2 + y_2 \equiv 0 (\bmod h_2), \quad 7m + 71 \equiv 0 (\bmod 6),$$
$$m - 1 \equiv 0 (\bmod 6) \quad m = 1, 7, 13 \cdots$$
$$m \equiv 1 (\bmod 6)$$

We must take $\underline{m_2 = 7} : (m_2, x_2, y_2, h_2) = (7, 7, 71, 6)$

Hence

$$x_3 = \frac{m_2 x_2 + y_2}{h_2} = 20,$$

$$y_3 = \frac{D x_2 + m_2 y_2}{h_2} = 203,$$

$$h_3 = \frac{D - m_2^2}{h_2} = 9$$

$$(x_3, y_3, h_3) = (20, 203, 9)$$

*Step 4:*

$$mx_3 + y_3 \equiv 0 (\bmod h_3),$$
$$20m + 203 \equiv 0 (\bmod 9),$$
$$2m + 5 \equiv 0 (\bmod 9)$$
$$m \equiv 2 (\bmod 9)$$

In the sequence $m = 2, 11, 20 \cdots$, we must take $\underline{m_3 = 2}$

$$(m_3, x_3, y_3, h_3) = (2, 20, 203, 9)$$

Hence

$$x_4 = \frac{m_3 x_3 + y_3}{h_3} = 27,$$

$$y_4 = \frac{Dx_3 + m_3y_3}{h_3} = 274,$$
$$h_4 = \frac{D - m_3^2}{h_3} = 11$$

$$(x_4, y_4, h_4) = (27, 274, 11)$$

*Step 5:*

$$\begin{aligned} mx_4 + y_4 &\equiv 0(\text{mod}\, h_4), \\ 27m + 274 &\equiv 0(\text{mod}\, 11), \\ 5m - 1 &\equiv 0(\text{mod}\, 11) \\ m &\equiv 9(\text{mod}\, 11) \end{aligned}$$

In the sequence $m = 9, 20, \cdots$, we must take $\underline{m_4 = 9}$
Hence

$$x_5 = \frac{m_4x_4 + y_4}{h_4} = 47,$$
$$y_5 = \frac{Dx_4 + m_4y_4}{h_4} = 477,$$
$$h_5 = \frac{D - m_4^2}{h_4} = 2$$

$$(x_5, y_5, h_5) = (47, 477, 2)$$

If we recall that $h_\lambda = (-1)^\lambda(Dx_\lambda^2 - y_\lambda^2)$, we have;

$$h_5 = 2 : Dx_5^2 - y_5^2 = -2$$

or

$$Dx_5^2 + 2 = y_5^2$$

By the lemma on composition

$$(\alpha, \beta) = (2x_5y_5, Dx_5^2 + y_5^2)$$

must satisfy the relation

$$D\alpha^2 + 4 = \beta^2$$

Hence follows that $(\alpha_0; \beta_0) = \left(\frac{\alpha}{2}, \frac{\beta}{2}\right)$ satisfies the relation

$$D\alpha_0^2 + 1 = \beta_0^2$$

Herein

$$\alpha_0 = \frac{\alpha}{2} = x_5y_5 = 22419$$

$$\beta_0 = \frac{\beta}{2} = \frac{1}{2}(Dx_5^2 + y_5^2) \quad = \frac{1}{2}(-2 + 2y_5^2)$$

$$= -1 + y_5^2 \qquad = 227528$$

$$(24) \qquad \begin{array}{|l} 103x^2 + 1 = y^2 \\ (x, y) = (22419, 227528) \end{array}$$

**Example 4.** [Narayana (1356)]

The equation

$$\boxed{Dx^2 + 1 = y^2 \text{ with } D = 97}$$

The application of Bhaskara's method, as described on p. 110, has been illustrated in detail in the previous examples: first the sequence of multipliers $m_\lambda$ is determined successively, along with the sequences $\{x_\lambda\}$, $\{y_\lambda\}$, $\{h_\lambda\}$. The aim is to arrive at $\lambda_0$ with $h_{\lambda_0} = 1$. In practice, as soon as we reach a $\lambda'$ with $h_{\lambda'} = \pm 2, \pm 4$ we can use Brahmagupta's lemmas to complete the solution.

Accordingly with $D = 97$, we have

$$9 < \sqrt{10} < 10, (x_0, y_0, h_0) = (1, 9, 16)$$

Then we find $m_0$ with

$$\begin{cases} m_0 x_0 + y_0 \equiv 0 (\mathrm{mod}\, h_0) \\ m_0 < \sqrt{D} < m_0 + h_0 \end{cases}$$

and find $m_0 = 7$, so that

$$(x_1, y_1, h_1) = (1, 10, 3)$$

And so on.

We shall not present all the details of the standard computations, but display the obtained results in the following tabular form

**D = 97**

| $\lambda$ | 0 | 1 | 2 | 3 | 4 | 5 | 6 | 7 | 8 | 9 | 10 |
|---|---|---|---|---|---|---|---|---|---|---|---|
| $m_\lambda$ | 7 | 8 | 3 | 5 | 4 | 5 | 3 | 8 | 7 | 9 | 9 |
| $x_\lambda$ | 1 | 1 | 6 | 7 | 13 | 20 | 33 | 53 | 86 | 483 | **569** |
| $y_\lambda$ | 9 | 10 | 59 | 69 | 128 | 197 | 325 | 522 | 847 | 4757 | **5604** |
| $h_\lambda$ | 16 | 3 | 11 | 8 | 9 | 9 | 8 | 11 | 3 | 16 | **1** |

Since we have here $h_{10} = 1, (x_{10}, y_{10}) = (569, 5604)$

$$Dx_{10}^2 - y_{10}^2 = 1 \text{ or } Dx_{10}^2 - 1 = y_{10}^2$$

Hence

$$(\alpha, \beta) = (2x_{10}y_{10}, Dx_{10}^2 + y_{10}^2)$$
$$= (2x_{10}y_{10}.1 + 2y_{10}^2)$$

must satisfy the relation $D\alpha^2 + 1 = \beta^2$.
In this manner

$$\boxed{\begin{aligned} &97x^2 + 1 = y^2 \\ &(x, y) = (6377352, 62809633) \end{aligned}} \quad (25)$$

## §4. HISTORICAL REMARKS

As we have already explained in several examples, the Chakravala method of Bhaskara consists in constructing, recursively, the sequences, (See p. 110),

$$(m_\lambda, x_\lambda, y_\lambda, h_\lambda) \quad (\lambda = 0, 1, 2, \cdots)$$

with the intial values:

$$1^0 \quad (x_0, y_0, h_0) = (1, [\sqrt{D}], F(x_0, y_0))$$
$$F(x, y) = Dx^2 - y^2$$
$$2^0 \quad m_0x_0 + y_0 \equiv 0 (\text{mod}\, h_0) \quad \text{with}$$
$$m_0 < \sqrt{D} < m_0 + h_0$$

Bhaskara (1150) in his BeejaGanitha has solved, among other examples, the equations

$$67x^2 + 1 = y^2 \text{ and } 61x^2 + 1 = y^2$$

It is remarkable that the equation

$$61x^2 + 1 = y^2$$

was proposed by Fermat to Frenicle in a letter of February, 1657; Euler solved it in 1732.

Narayana (1356) in his Ganita-Kaumudi has solved the equations

$$97x^2 + 1 = y^2 \text{ and } 103x^2 + 1 = y^2$$

The problem of determining *integer* solutions of algebraic equations is called ≪DIOPHANTINE ANALYSIS≫ after the Greek mathematician Diophantus of Alexandria at about the third century (A.D.). A part of the work ≪Arithmetica≫ of Diophantus was studied by Fermat. It was Fermat who initiated its study in Europe.

Fermat, in one of his well-known challenges to the British mathematicians of his time proposed the equation

$$(1) \qquad \boxed{Dx^2 + 1 = y^2,} \quad D > 0 \text{ integer } \neq \text{ a square.}$$

It was an accident that Euler referred to it as the ≪ PELL EQUATION ≫ but it has no historical justification whatever. There is no contribution by Pell to this topic.

The obvious solution $(x, y) = (0, 1)$ of the equation (1) is called its trivial solution. As soon as one finds a non-trivial solution $(x_0, y_0)$ an infinite number of new solutions can be found by repeated application of the Principle of compositions (Brahmagupta's Bhavana Principle).

It is Bhaskara's Chakravala method that makes the decisive step in determining a *non-trivial* solution.

Under these circumstances it should be appropriate to designate the equation (1) as the *Brahmagupta-Bhaskara equation* or even as the ≪ BHASKARA EQUATION ≫ since it is his chakravalam method that makes the crucial step.

Bhaskara's CHAKRAVALA method has a very intimate connection with the continued fraction algorithm.

In the following section we gather some basic facts regarding continued fractions. However, for its own interest, we shall present the topic in slightly greater detail than what is absolutely needed for our immediate purpose.

## § 5. CONTINUED FRACTIONS

Let $\{a_n\}$, $(n = 0, 1, 2, \cdots)$, be a sequence of *integers* such that

$$a_0 \geq 0, a_n \geq 1 \;\; (n = 1, 2, \cdots)$$

Then

(1)
$$a_0 + \cfrac{1}{a_1 + \cfrac{1}{a_2 + \ddots + \cfrac{1}{a_n + \ddots}}}$$

is called a ≪ *simple continued fraction* ≫. We shall only deal with *simple* continued fractions and *not* with continued fractions of more general type

(2)
$$a_0 + \cfrac{b_1}{a_1 + \cfrac{b_2}{a_2 + \ddots + \cfrac{b_n}{a_n + \ddots}}}$$

Accordingly, while referring to (1), we often omit the word ≪ *simple* ≫ and speak of merely continued fractions.

*NOTATION:* In order to save space we find it convenient to write (1) as

(3)
$$C_n = a_0 + \frac{1}{a_1 +} \; \frac{1}{a_2 + \cdots} \; \frac{1}{+a_n}$$

or also as

(4)
$$C_n = [a_0, a_1, a_2, \cdots, a_n]$$

Thus

$$C_0 = a_0, C_1 = a_0 + \frac{1}{a_1} = \frac{a_0 a_1 + 1}{a_1},$$

$$C_2 = a_0 + \cfrac{1}{a_1 + \cfrac{1}{a_2}} = \frac{a_2 a_1 a_0 + a_2 + a_0}{a_2 a_1 + 1}$$

**Theorem 1.**

$1^0$ **Define** the sequences $p_n$ and $q_n$, $(n = 0, 1, 2, \ldots)$, by means of the recurrence relations

(5)
$$\boxed{\begin{aligned} p_n &= a_n p_{n-1} + p_{n-2} \\ q_n &= a_n q_{n-1} + q_{n-2} \end{aligned}} \qquad (n = 2, 3, \cdots)$$

with the initial values

(6)
$$\begin{pmatrix} p_0 & p_1 \\ q_0 & q_1 \end{pmatrix} = \begin{pmatrix} a_0 & a_0 a_1 + 1 \\ 1 & a_1 \end{pmatrix}$$

Then the $n$-th convergents $C_n$ are given by

$$(*)\qquad \boxed{C_n = a_0 + \frac{1}{a_1+}\cdots\frac{1}{+a_n} = \frac{p_n}{q_n}}$$

$2^0$ The convergents satisfy the relations

$$(7)\qquad \boxed{p_n q_{n-1} - p_{n-1} q_n = (-1)^{n-1}}$$

**Proof:** We shall prove the assertion (*) by induction on $n$. For $n = 0$, we have $C_0 = a_0$, $p_0 = a_0$, $q_0 = 1$, (see (6)) thus $C_0 = \dfrac{p_0}{q_0}$. Again, for $n = 1$, we have

$$C_1 = a_0 + \frac{1}{a_1},\ \ p_1 = a_0 a_1 + 1,\ \ q_1 = a_1,\ \text{(see (6))}$$

so that $C_1 = \dfrac{p_1}{q_1}$. The assertion (*) holds for $n = 0, 1$. Assume that $n \geq 2$ and $C_n = \dfrac{p_n}{q_n}$.

It is clear from the recurrence relations that $p_n$ is a polynomial in $a_0, a_1, \cdots a_n$ and $q_n$ is a polynomial in $a_1, \cdots a_n$. Now, by inductive assumption

$$C_n = \frac{p_n}{q_n} = \frac{a_n p_{n-1} + p_{n-2}}{a_n q_{n-1} + q_{n-2}}$$

since $C_{n+1}$ results from $C_n$ by replacing $a_n$ by $a_n + \dfrac{1}{a_{n+1}}$ and $p_{n-1}$, $p_{n-2}$, $q_{n-1}$, $q_{n-2}$ are independent of $a_n$, we get

$$C_{n+1} = \frac{\left(a_n + \dfrac{1}{a_{n+1}}\right) p_{n-1} + p_{n-2}}{\left(a_n + \dfrac{1}{a_{n+1}}\right) q_{n-1} + q_{n-2}}$$

In view of the recursive relations (5) this becomes

$$C_{n+1} = \frac{(a_n p_{n-1} + p_{n-2}) + \dfrac{1}{a_{n+1}} p_{n-1}}{(a_n q_{n-1} + q_{n-2}) + \dfrac{1}{a_{n+1}} q_{n-1}} = \frac{p_n + \dfrac{1}{a_{n+1}} p_{n-1}}{q_n + \dfrac{1}{a_{n+1}} q_{n-1}} =$$

$$= \frac{a_{n+1} p_n + p_{n-1}}{a_{n+1} q_n + q_{n-1}} = \frac{p_{n+1}}{q_{n+1}},$$

which completes the induction and establishes the assertion (*).

$2^0$ From the recurrence relations (5) we have:

$$p_n \cdot q_{n-1} - p_{n-1} \cdot q_n = (a_n p_{n-1} + p_{n-2})q_{n-1} - p_{n-1}(a_n q_{n-1} + q_{n-2})$$
$$= -(p_{n-1}q_{n-2} - p_{n-2}q_{n-1})$$

Hence, if we put, $r_n = p_n q_{n-1} - p_{n-1}q_n$, we have

$$r_n = -r_{n-1}$$

Accordingly

$$r_n = -r_{n-1} = +r_{n-2} = \cdots = (-1)^k r_{n-k} = \cdots = (-1)^{n-1} \cdot r_1$$

while $r_1 = p_1 q_0 - p_0 q_1 = (a_0 a_1 + 1) \cdot 1 - a_0 \cdot a_1 = 1$. This proves $2^0$ and $3^0$ is an immediate consequence of the relation $p_n q_{n-1} - p_{n-1} q_n = (-1)^{n-1}$.

*Remark* Since all the $a_n$, for $n \geq 1$, are assumed to be positive integers, thus $a_n \geq 1$ for $n = 1, 2, \cdots$, it follows from (5) that

(8) $$q_n \geq a_n q_{n-1} + 1 \geq q_{n-1} + 1$$

and consequently

(9) $$\boxed{q_n \geq n}$$

**Continued fraction Algorithm for any real number**

Let $\theta$ be any real number. Denote by $t_0$ the integral part of $\theta$: thus

$$t_0 = [\theta], \quad t_0 < \theta < t_0 + 1$$

(We assume here that $\theta$ is an *irrational* number. For rational numbers one gets finite simple continued fractions which have already been considered earlier) [See Chapter I, § 4 pp 18-24]. From (10) the fractional part $\theta - t_0$ is such that

$$0 < \theta - t_0 < 1$$

so that

$$\frac{1}{\theta - t_0} > 1$$

Denote by $\theta_1'$ the reciprocal of the fractional part

$$\theta - [\theta] = \theta - t_0$$

i.e.,

(11) $$\theta_1' = \frac{1}{\theta - t_0} > 1$$

In this manner

(12) $$\boxed{\theta = t_0 + \frac{1}{\theta_1'}} \quad \theta_1' > 1$$

This is the first step of the continued fraction algorithm. We now proceed with $\theta'$, as with $\theta$. We obtain an infinite sequence of *integers*

$$t_0, t_1, t_2, \cdots t_n, \cdots$$

such that

$$\text{(13)} \qquad \left. \begin{aligned} \theta &= t_0 + \frac{1}{\theta'_1} \\ \theta'_1 &= t_1 + \frac{1}{\theta'_2} \\ &\vdots \\ \theta'_n &= t_n + \frac{1}{\theta'_{n+1}} \\ &\vdots \end{aligned} \right| \quad \begin{aligned} t_0 &= [\theta] \geq 0 \\ t_n &= [\theta'_n] \geq 1 \\ (n &= 1, 2, \cdots) \end{aligned}$$

We call $\theta'_n$ the $(n+1)^{th}$ *complete quotient* of the continued fraction

$$[t_0, t_1, \cdots, t_n, \cdots]$$

The process does not terminate since, otherwise, $\theta$ would turn out to be rational against our hypothesis.

We need the following lemma which will be often used.

**LEMMA 1**

The $n$th convergents,

$$C_n = \frac{p_n}{q_n} = [t_0, t_1, \cdots, t_n]$$

and the $(n+1)$th complete quotients

$$\theta'_n = [t_n, t_{n+1}, \cdots]$$

of the continued fraction

$$\theta = [t_0, t_1, \cdots, t_n, \cdots]$$

are related by the relations

(14)

$$\boxed{\theta = \theta'_0 = \frac{\theta'_1 t_0 + 1}{\theta'_1} = \frac{\theta'_2 p_1 + p_0}{\theta'_2 q_1 + q_0} = \cdots = \frac{\theta'_n p_{n-1} + p_{n-2}}{\theta'_n q_{n-1} + q_{n-2}} = \cdots}$$

**Proof:** This follows at once from the defining relations (13). At the outset,

$$\theta = t_0 + \frac{1}{\theta'_1} = \frac{\theta'_1 t_0 + 1}{\theta'_1} = \frac{\left(t_1 + \frac{1}{\theta'_2}\right) t_0 + 1}{t_1 + \frac{1}{\theta'_2}}$$

$$= \frac{p_1 + \frac{t_0}{\theta'_2}}{q_1 + \frac{1}{\theta'_2}} = \frac{\theta'_2 p_1 + p_0}{\theta'_2 q_1 + q_0},$$

since [see (6), P. 122]

$$\begin{pmatrix} p_0 & p_1 \\ q_0 & q_1 \end{pmatrix} = \begin{pmatrix} t_0 & t_0 t_1 + 1 \\ 1 & t_1 \end{pmatrix}$$

Using (5) [See P. 122] we can now prove (14) by induction on $n \geq 2$, i.e.,

$$\theta = \frac{\theta'_n p_{n-1} + p_{n-2}}{\theta'_n q_{n-1} + q_{n-2}} \quad (n \geq 2)$$

In fact, assume that (15) holds for $n$. For $n = 2$ we have verified it above. Since $\theta'_n = t_n + \dfrac{1}{\theta'_{n+1}}$ we have from (15)

$$\frac{\theta'_n p_{n-1} + p_{n-2}}{\theta'_n q_{n-1} + q_{n-2}} = \frac{\left(t_n + \frac{1}{\theta'_{n+1}}\right) p_{n-1} + p_{n-2}}{\left(t_n + \frac{1}{\theta'_{n+1}}\right) q_{n-1} + q_{n-2}}$$

$$= \frac{(t_n p_{n-1} + p_{n-2}) + \frac{p_{n-1}}{\theta'_{n+1}}}{(t_n q_{n-1} + q_{n-2}) + \frac{q_{n-1}}{\theta'_{n+1}}}$$

$$= \frac{p_n + \frac{p_{n-1}}{\theta'_{n+1}}}{q_n + \frac{q_{n-1}}{\theta'_{n+1}}}$$

$$= \frac{\theta'_{n+1} p_n + p_{n_1}}{\theta'_{n+1} q_n + q_{n_1}},$$

which, because of (15), gives

$$\theta = \frac{\theta'_{n+1} p_n + p_{n_1}}{\theta'_{n+1} q_n + q_{n_1}},$$

i.e. (15) holds for $n + 1$. The induction is complete and the lemma is proved.

**Theorem 2.**

*Every irrational number has a uniquely determined simple continued fraction expansion*

**Proof:** Let

$$\theta = \begin{vmatrix} [t_0, t_1, t_2, \cdots] \\ \\ [a_0, a_1, a_2, \cdots] \end{vmatrix}$$

We have $t_0 = [\theta] = a_0$. In the chain of equations

$$\begin{array}{l|l} \theta = t_0 + \dfrac{1}{\theta'_1} & \theta = a_0 + \dfrac{1}{\alpha'_1} \\ \theta'_1 = t_1 + \dfrac{1}{\theta'_2} & \alpha'_1 = a_1 + \dfrac{1}{\alpha'_2} \\ \vdots & \vdots \\ \theta'_n = t_n + \dfrac{1}{\theta'_{n+1}} & \alpha'_n = a_n + \dfrac{1}{\alpha'_{n+1}} \\ \vdots & \vdots \\ \theta'_n = [t_n, t_{n+1}, \cdots] & \alpha'_n = [a_n, a_{n+1}] \end{array}$$

We find that

$$t_0 + \frac{1}{\theta_,'} = \theta = a_0 + \frac{1}{\alpha'_1} \text{ so that } \frac{1}{\theta'_1} = \frac{1}{\alpha'_1} \text{ or } \theta'_1 = \alpha'_1$$

Hence $t_1 = [\theta'_1] = [\alpha'_1] = a_1$, then from the next equation we see that $t_2 = a_2$ and so on. We prove by induction that $t_n = a_n$ for $n = 0, 1, 2, \cdots$. Since,

$$\theta = \begin{vmatrix} [t_0, t_1, \cdots, t_{n-1}, \theta'_n] \\ \\ [a_0, a_1, \cdots, a_{n-1}, \alpha'_n], \end{vmatrix}$$

Assume, as hypothesis of induction, that

$$a_k = t_k (k = 0, 1, \cdots, n-1)$$

Then, from

$$\theta = [t_0, \cdots, t_{n-1}, \theta'_n] = [t_0, \cdots, t_{n-1}, \alpha'_n]$$

we find

$$\theta = \frac{\theta'_n p_{n-1} + p_{n-2}}{\theta'_n q_{n-1} + q_{n-2}} = \frac{\alpha'_n p_{n+1} + p_{n-2}}{\alpha'_n q_{n-1} + q_{n-2}} \tag{16}$$

where

$$\frac{p_n}{q_n} = [t_0, \cdots, t_n]$$

[See Lemma 1, P. 125]. From (16) we find,

$$(\theta'_n p_{n-1}+p_{n-2})(\alpha'_n q_{n-1}+q_{n-2})=(\theta'_n q_{n-1}+q_{n-2})(\alpha'_n p_{n-1}+p_{n-2})$$

$$\theta'_n(p_{n-1}q_{n-2}-q_{n-1}p_{n-2})=\alpha'_n(p_{n-1}q_{n-2}-q_{n-1}p_{n-2})$$

From (7), $p_{n-1}q_{n-2}-q_{n-1}p_{n-2}=(-1)^{n-2}$, therefore $\theta'_n=\alpha'_n$,
Consequently $t_n=[\theta'_n]=[\alpha'_n]=a_n$.■

**Theorem 3.**

*If $\theta$ is irrational, and $C_n$ is its $n^{th}$ convergent*

$$\lim_{n\to\infty} C_n=\theta$$

**Proof:** We have from (14):

$$C_n=\frac{p_n}{q_n} \text{ and } \theta=\frac{\theta'_{n+1}p_n+p_{n-1}}{\theta'_{n+1}q_n+q_{n-1}}$$

so that

$$\theta-C_n=\frac{\theta'_{n+1}p_n+p_{n-1}}{\theta'_{n+1}q_n+q_{n-1}}-\frac{p_n}{q_n}$$

$$=\frac{p_{n-1}q_n-p_nq_{n-1}}{q_n(\theta'_{n+1}q_n+q_{n-1})}$$

(17) $$=\frac{-(-1)^{n-1}}{q_n(\theta'_{n+1}q_n+q_{n-1})}$$

Since $\theta'_{n+1}=t_{n+1}+\dfrac{1}{\theta'_{n+2}}>t_{n+1}$, thus $\theta'_{n+1}q_n+q_{n-1}>t_{n+1}q_n+q_{n-1}=q_{n+1}$

We have

$$0<\frac{1}{\theta'_{n+1}q_n+q_{n-1}}<\frac{1}{q_{n+1}}$$

Hence from (17) follows

(18) $$\boxed{|\theta-C_n|<\frac{1}{q_nq_{n+1}}} \quad (n<1,2,\cdots)$$

From (9), [See P. 121], $q_n\geq n$, thus

$$q_nq_{n+1}\geq n(n+1)\longrightarrow\infty$$

Consequently the assertion of the theorem follows from (18).

**Example 1.** Continued fraction for $\sqrt{D}$ : $\underline{D=61}$. Using the relation

$$\sqrt{D}-a=\frac{D-a^2}{\sqrt{D}+a}$$

repeatedly, and separating the integral parts at each step we can easily obtain the continued fraction expansion of $\sqrt{D}$, i.e. we put $a_0 = [\sqrt{D}]$ and write

$$\sqrt{D} = a_0 + (\sqrt{D} - a_0) = a_0 + \frac{D - a_0^2}{\sqrt{D} + a_0} = a_0 + \frac{1}{\delta_1},$$

where

$$\delta_1 = \frac{\sqrt{D} + a_0}{D - a_0^2} = a_1 + \frac{1}{\delta_2}, a_1 = [\delta_1]$$

And so on. Accordingly

$$\sqrt{61} = 7 + \underline{(\sqrt{61} - 7)} = \mathbf{7} + \frac{\mathbf{12}}{\sqrt{\mathbf{61}} + \mathbf{7}}$$

$$\frac{\sqrt{61} + 7}{12} = 1 + \frac{\sqrt{61} - 5}{12} = 1 + \frac{3}{\sqrt{61} + 5}$$

$$\frac{\sqrt{61} + 5}{3} = 4 + \frac{\sqrt{61} - 7}{3} = 4 + \frac{4}{\sqrt{61} + 7}$$

$$\frac{\sqrt{61} + 7}{4} = 3 + \frac{\sqrt{61} - 5}{4} = 3 + \frac{9}{\sqrt{61} + 5}$$

$$\frac{\sqrt{61} + 5}{9} = 1 + \frac{\sqrt{61} - 4}{9} = 1 + \frac{5}{\sqrt{61} + 4}$$

$$\frac{\sqrt{61} + 4}{5} = 2 + \frac{\sqrt{61} - 6}{5} = 2 + \frac{5}{\sqrt{61} + 6}$$

$$\frac{\sqrt{61} + 6}{5} = 2 + \frac{\sqrt{61} - 4}{5} = 2 + \frac{9}{\sqrt{61} + 4} \tag{19}$$

$$\frac{\sqrt{61} + 4}{9} = 1 + \frac{\sqrt{61} - 5}{9} = 1 + \frac{4}{\sqrt{61} + 5}$$

$$\frac{\sqrt{61} + 5}{4} = 3 + \frac{\sqrt{61} - 7}{4} = 3 + \frac{3}{\sqrt{61} + 7}$$

$$\frac{\sqrt{61} + 7}{3} = 4 + \frac{\sqrt{61} - 5}{3} = 4 + \frac{12}{\sqrt{61} + 5}$$

$$\frac{\sqrt{61} + 5}{12} = 1 + \frac{\sqrt{61} - 7}{12} = 1 + \frac{1}{\sqrt{61} + 5}$$

$$\frac{\sqrt{61} + 7}{1} = 14 + \underline{(\sqrt{\mathbf{61}} - \mathbf{7})}$$

We are back at the first step in (19). The process gets indefinitely repeated. We say that the simple continued fraction expansion of $\sqrt{D}$ is *periodic*. We agree to write the expansion as

$$\boxed{\sqrt{61} = [7; \overset{*}{1}, 4, 3, 1, 2, 2, 1, 3, 4, 1, \overset{*}{14}]}$$

The asteriks mean that the sequence

$$1, 4, 3, 1, 2, 2, 1, 3, 4, 1, 14$$

recur indefinitely.

The sequence of convergents $C_n = \frac{p_n}{q_n}$ can be written down by means of the recurrence relations (5) in Theorem 1 [See p. 122]. It can be conveniently displayed in a tabular form.

(21)

| $n$ | 0 | 1 | 2 | 3 | 4 | 5 | 6 | 7 | 8 | 9 | 10 | 11 |
|---|---|---|---|---|---|---|---|---|---|---|---|---|
| $a_n$ | 7 | 1 | 4 | 3 | 1 | 2 | 2 | 1 | 3 | 4 | 1 | 14 |
| $p_n$ | 7 | 8 | 39 | 125 | 164 | 453 | 1070 | 1523 | 5639 | 24079 | **29718** | — |
| $q_n$ | 1 | 1 | 5 | 16 | 21 | 58 | 137 | 195 | 722 | 3083 | **3805** | — |

We have not entered the terms $p_n$ and $q_n$ for $n = 11$ since we do not need them in what follows. We recall, [See (22'), p. 115], that with $D = 61$,

$$\text{(22)} \quad \boxed{\begin{aligned} Dx^2 - 1 = y^2 \text{ for } (x, y) &= (3805, 29718) \\ &= (q_{10}, p_{10}) \end{aligned}}$$

We have seen in Theorem 3 that $\frac{p_n}{q_n} = C_n$ converges to $\theta$ very fast, and offes a good approximation to $\theta$ [See (18), P. 128]. But very soon the denominators $q_n$ become extremely large. Hence the question arises whether we can find a rational approximation $\frac{a}{b}$ to $\theta$, with $0 < b < q_n$, which approximates better than $\frac{p_n}{q_n}$, i.e. *is it possible that*

$$\left|\theta - \frac{a}{b}\right| < \left|\theta - \frac{p_n}{q_n}\right| \text{ with } 0 < b < q_n?$$

It turns out that this is not possible.

**Theorem 4. (The law of best approximation)**

*If $\theta > 0$ is irrational, and $C_n = \frac{p_n}{q_n}$ $(n = 0, 1, 2, \ldots)$ is the* *sequence of convergents to $\theta$, then for all rational numbers*

$$\frac{a}{b} \text{ with } 0 < b < q_n$$

*there holds the inequality.*

(23) $$\boxed{|\theta - C_n| < \left|\theta - \frac{a}{b}\right|}$$

**Proof**: First we observe that

(24) $$\boxed{p_n q_{n-2} - p_{n-2} q_n = (-1)^n a_n} \qquad (n \geq 2)$$

In fact,

$$\begin{aligned} p_n q_{n-2} - p_{n-2} q_n &= (a_n p_{n-1} + p_{n-2}) q_{n-2} - p_{n-2}(a_n q_{n-1} + q_{n-2}) \\ &= a_n (p_{n-1} q_{n-2} - p_{n-2} q_{n-1}) \\ &= a_n.(-1)^{n-2} = a_n.(-1)^n \end{aligned}$$

since $p_n q_{n-1} - p_{n-1} - p_{n-1} q_n = (-1)^{n-1}$. This proves (24).

From (24) we obtain

$$C_n - C_{n-2} = \frac{p_n}{q_n} - \frac{p_{n-2}}{q_{n-2}} = (-1)^n . \frac{a_n}{q_n q_{n-2}}$$

whence follows

$$C_{2k} - C_{2k-2} = \frac{a_{2k}}{q_{2k} q_{2k-2}} > 0$$

$$C_{2k-1} - C_{2k-3} = -\frac{a_{2k-1}}{q_{2k-1} q_{2k-3}} < 0$$

Again since

$$C_n - C_{n-1} = \frac{p_n}{q_n} - \frac{p_{n-1}}{q_{n-1}} = \frac{(-1)^{n-1}}{q_n q_{n-1}}$$

we get

$$C_{2k} - C_{2k-1} < 0$$

In other words $\{C_{2k}\}$ is *increasing* while $\{C_{2k-1}\}$ is *decreasing*, and $C_{2k} < C_{2k-1}$:

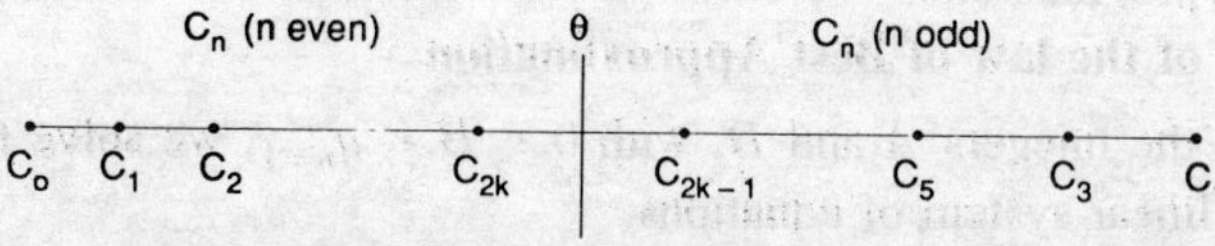

From (14) we have

$$\theta = \frac{\theta'_{n+1}p_n + p_{n-1}}{\theta'_{n+1}q_n + q_{n-1}} \text{ whence follows } \theta'_{n+1} = -\frac{q_{n-1}\theta - p_{n-1}}{q_n\theta - p_n}$$

If we put

$$\boxed{r_n = (-1)^n.(q_n\theta - p_n)}$$

we can write

(25) $$\theta'_{n+1} = -\frac{(-1)^{n-1}r_{n-1}}{(-1)^n r_n} = \frac{r_{n-1}}{r_n}$$

Since $\theta'_{n+1} > 1$ we conclude that $r_n$ has the same sign as $r_{n-1}$. Since $r_0 = q_0\theta - p_0 = \theta - t_0 = \theta - [\theta] > 0$, it follows that, successively,

$$r_0 > 0, r_1 > 0, r_2 > 0, \ldots, r_n > 0 \ldots$$

and then from (25) we see that

$$r_{n-1} > r_n$$

i.e.

(26) $$\frac{r_n \text{ is a decreasing sequence of positive integers}}{r_0 > r_1 > \cdots > r_{n-1} > r_n > \cdots > 0}$$

Since, always, $q_n > q_{n-1}$, thus $\dfrac{1}{q_{n-1}} > \dfrac{1}{q_n}$, (26) gives

(27) $$\frac{r_{n-1}}{q_{n-1}} > \frac{r_n}{q_n}$$

But $\dfrac{r_n}{q_n} = (-1)^n(\theta - C_n)$, and since $r_n > 0$,

$$|\theta - C_n| = \frac{r_n}{q_n}$$

Consequently (27) means that

(28) $$\boxed{|\theta - C_n| < |\theta - C_{n-1}|}$$

i.e.

> In the sequence of convergents $C_0, C_1, \ldots, C_{n-1}, C_n, \ldots$ of $\theta$, each term is a better approximation to $\theta$ then the previous one.

**Proof of the law of Best Approximation**

Given the integers $A$ and $B$, with $0 < B < q_{n+1}$, we solve for $\alpha$ and $\beta$ the linear system of equations

$$\begin{cases} A &= \alpha p_n + \beta p_{n+1} \\ B &= \alpha q_n + \beta q_{n+1} \end{cases}$$

since the determinant of this system is

$$\begin{vmatrix} p_n & p_{n+1} \\ q_n & q_{n+1} \end{vmatrix} = (-1)^{n-1}$$

it follows that $\alpha$ and $\beta$ must be *integers*. First, we observe that $\alpha \neq 0$: for, otherwise, $\alpha = 0 \Rightarrow B = \beta q_{n+1} \Rightarrow 0 < \beta < 1$, since $0 < B < q_{n+1}$. This is impossible since $\beta$ is an integer

Next if $\beta \neq 0$, we claim that $\alpha$ and $\beta$ must be of opposite sign. In fact

$$\alpha < 0, \quad \beta < 0 \Rightarrow B = \alpha q_n + \beta q_{n+1} < 0$$

$$\alpha > 0, \quad \beta > 0 \Rightarrow B > \beta q_{n+1} \geq q_{n+1}$$

This is impossible since $0 < B < q_{n+1}$.

Consequently,

$\alpha \neq 0$, and if $\beta \neq 0$, then $\alpha$ and $\beta$ are of *opposite* sign

We have

$$\begin{aligned} B\theta - A &= (\alpha q_n + \beta q_{n+1})\theta - (\alpha p_n + \beta p_{n+1}) \\ &= \alpha(q_n\theta - p_n) + \beta(q_{n+1}\theta - p_{n+1}) \\ &= \alpha \cdot (-1)^n r_n + \beta \cdot (-1)^{n+1} \cdot r_{n+1} \\ &= (-1)^n \cdot (\alpha r_n - \beta) \end{aligned}$$

Since $r_n > 0$, and $\alpha$ and $-\beta$ are of the *same* sign we conclude:

$$|B\theta - A| = |\alpha r_n - \beta| = |\alpha|\, r_n + |\beta| \geq |\alpha|\, r_n \geq r_n$$

In this manner

$$0 < B < q_{n+1} \Rightarrow |B\theta - A| \geq r_n = |q_n\theta - p_n| \tag{29}$$

We shall deduce the Law of Best Approximation, namely the inequality (23), (see p. 131), from (29): Suppose now that $0 < b < q_n$. Since, always, $q_n < q_{n+1}$ we have $0 < b < q_{n+1}$. We obtain from (29) that

$$|b\theta - a| \geq |q_n\theta - p_n|$$

Accordingly,

$$\begin{aligned} \left|\theta - \frac{p_n}{q_n}\right| = \frac{1}{q_n}|q_n\theta - p_n| &\leq \frac{1}{q_n}|b\theta - a| \\ &\leq \frac{1}{b}|b\theta - a| = \left|\theta - \frac{a}{b}\right| \end{aligned}$$

i.e.

$$|\theta - C_n| < \left|\theta - \frac{a}{b}\right|,$$

which yields the required estimate (23).

The proof of Theorem 4 is hereby complete ■

The following remarkable fact is an immediate consequence of the Law of Best Approximation:

**Theorem 5.**

*If $\theta$ is an irrational number, and $\frac{a}{b}$, $(b > 0)$, is a rational approximation to $\theta$ satisfying the inequality*

$$\boxed{\left|\theta - \frac{a}{b}\right| < \frac{1}{2b^2}} \tag{30}$$

*then $\frac{a}{b}$ must be a convergent to $\theta$, i.e. there exists a uniquely determined index $n_0$ such that*

$$\boxed{\frac{a}{b} = \frac{p_{n_0}}{q_{n_0}}} \tag{31}$$

**Proof**: Since $q_n$ is a strictly monotonic increasing sequence of positive integers, we can *uniquely* fix the index $n_0$ such that

$$\boxed{0 < q_{n_0} \leq b < q_{n_0+1}}$$

In view of (29) we therefore have:

$$\left|q_{n_0}\theta - p_{n_0}\right| \leq |b\theta - a| \text{ and } q_{n_0} \leq b \tag{32}$$

We can write

$$aq_{n_0} - bp_{n_0} = b(q_{n_0}\theta - p_{n_0}) + q_{n_0}(a - b\theta)$$

so that

$$\begin{aligned} \left|aq_{n_0} - bp_{n_0}\right| &\leq b \cdot \left|q_{n_0} - p_{n_0}\right| + q_{n_0}\left|a - b\theta\right| \\ &\leq b \cdot |b\theta - a| + b \cdot |b\theta - a|, && \textit{See}(32) \\ &= 2b \cdot |b\theta - a| \\ &= 2b^2\left|\theta - \frac{a}{b}\right| < 1, && \textit{See } (30) \end{aligned}$$

Hence we obtain that

$$\left|aq_{n_0} - bp_{n_0}\right| < 1$$

Here on the left side there is a *non-negative integer*. Consequently we must have

$$aq_{n_0} - bp_{n_0} = 0, \text{ i.e. } \frac{a}{b} = \frac{p_{n_0}}{q_{n_o}}$$

which establishes (31) and completes the proof of the theorem ■

## § 6. REMARKS ON $\pi$

It is known that the circumference of a circle bears a constant ratio to its diameter: this constant is denoted by $\pi$:

$$\boxed{\pi = \frac{\text{circumference of a circle}}{\text{its diameter}}}$$

A number $\alpha$ is said to be *algebraic* if it is a root of a polynomial equation with *integer* coefficients. i.e. it satisfies a relation of the form

$$a_0\alpha^n + a_1\alpha^{n-1} + \ldots + a_n = 0$$

$$a_0, a_1, \ldots, a_n \quad \textit{integers}$$

If no such relation holds for $\alpha$, then $\alpha$ is said to be *transcendental*.

We state, without proof, the following:

**Theorem** (LINDEMANN)

*The number $\pi$ is transcendental.*

In ≪*Aryabhatiya, Section 2: Ganitapada*≫ a rational approximation to $\pi$ is given:

चतुरधिकं शतमष्टगुणं द्वाषष्टिस्तथा सहस्राणाम् ।
अयुतद्वयविष्कम्भस्यासन्नो वृत्तपरिणाहः ।।

Circumference of a circle with diameter 20000

$$= 62832$$

i.e.

$$\pi \simeq \frac{62832}{20000} = 3.1416 \tag{32}$$

Nilakantha Somayajin, in his commentary, tries to explain that $\pi$ can never be equal to a rational number.

In ≪*Kriyakramakari*≫ $\frac{355}{113}$ is given as a rational approximation to $\pi$.

Nilakantha Somayajin credits Madhava with the rational approximation

$$\begin{aligned}(33) \qquad \pi &\simeq \frac{2\,827\,433\,388\,233}{9\,000\,000\,000\,000}\\ &= 3.1415\,9265\,359\end{aligned}$$

In ≪*Tantrasamgraha*≫ quotation in ≪YUKTIBHASHA≫ an infinite series expansion has been described for $\pi$. C. T. Rajagopal and Mukunda Marar have described such series, along with the original verses, as an appendix to their memoir ≪On the Hindu quadrature of the circle≫.

व्यासे वारिधिनिहते रूपहते व्याससागराभिहते ।
त्रिशरादिविषमसंख्या भक्तं ऋणं स्वं पृथक्क्रमात् कुर्यात् ।।

i.e.

$$(34) \qquad \boxed{\text{circumference } = 4d - \frac{4d}{3} + \frac{4d}{5} - \frac{4d}{7} + \ldots}$$

The following verse describes in the K a t a p y ā d i system a rational approximation to $\pi$

| च | न्डां | शु | चं | द्रा | ध | म | कुं | भि | पा | ल |
|---|---|---|---|---|---|---|---|---|---|---|
| 6 | 3 | 5 | 6 | 2 | 9 | 5 | 1 | 4 | 1 | 3 |

for a radius of आ नू न नू न्ना न न नु त्र नि त्य म्
i.e.,

$$(35) \qquad \boxed{\frac{\pi}{10} \simeq \cdot 3141\ \ 5926\ \ 5926\ \ 536}$$

or

$$\pi \simeq \frac{31\,415\,926\,536}{10\,000\,000\,000}$$

*Comments on the Katapayadi system,*

कादि नव, टादि नव, पादि पञ्चक, याद्यष्टक

| 1 | 2 | 3 | 4 | 5 | 6 | 7 | 8 | 9 |
|---|---|---|---|---|---|---|---|---|
| क | ख | ग | घ | ङ | च | छ | ज | झ |
| ट | ठ | ड | ढ | ण | त | थ | द | ध |
| प | फ | ब | भ | म | | | | |
| य | र | ल | व | श | ष | स | ह | क्ष-० |

Sri Bharathi Krishna Thirtha in his ≪VEDIC MATHEMATICS≫ gives another verse:

गोपी भाग्य मधुव्रात श्रृङ्गिशो दधिसन्धिग ।
खलजीवित खाताव गलहालारसंधर ।

(36) $$\frac{\pi}{10} = 0.3\ 1415\quad 9265\quad 3589\quad 7932\quad 3846\quad 2643\quad 3832\quad 79\ldots$$

The flexibility of Sanskrit makes the construction of such stanzas a routine task.

There is a verse in French, quoted by E. Cesáro, which gives the rational approximation for $\pi$ given in (36):

(37)
*Que j'aime à faire apprendre un nombre utile aux sages!*
*Immortel Archimede, artiste ingénieur*
*Qui de ton jugement peut priser la valeur?*
*Pour moi ton problème eut de pareils avantages.*

We count the number of letters in each word of the stanza (37) in order to obtain the rational approximation (36), namely

(36') $$\boxed{\pi = 3\cdot\ 1415926535|\ 8979|\ 32384626|43383279|\cdots}$$

(correct to 30 decimal places). [Compare (36)]

The series (34) for $\pi$ given by

$$\text{(38)} \qquad \boxed{\frac{\pi}{4} = 1 - \frac{1}{3} + \frac{1}{5} - + \ldots}$$

is not very convenient for obtaining rational aproximations to $\pi$. It is necesary to take *24000* terms to obtain the value of $\pi$ correct to just 4 decimal places:

$$\pi = 3.1415\ldots$$

Using the series

$$\tan^{-1} x = x - \frac{x^3}{3} + \frac{x^5}{5} - + \cdots (-1 < x \leq 1)$$

and putting $\tan\alpha = \dfrac{1}{5}$ we have:

$$\tan 2\alpha = \frac{5}{12}, \quad \tan 4\alpha = 1 + \frac{1}{119}$$

Let $\beta = 4\alpha - \dfrac{\pi}{4}$, then $\tan\beta = \dfrac{1}{239}$. Since

$$\text{(39)} \qquad \frac{\pi}{4} = 4\alpha - \beta = 4\tan^{-1}\frac{1}{5} - \tan^{-1}\frac{1}{239}$$

we get

$$\text{(40)} \qquad \pi = \frac{16}{5}\left[1 - \frac{4}{3.10^2} + \frac{4^2}{5.10^4} - \frac{4^3}{7.10^6} + - \ldots\right] - \\ - \frac{4}{239}\left[1 - \frac{1}{3.\delta} + \frac{1}{5.\delta^2} - \frac{1}{7.\delta^3} + - \ldots\right]$$

wherein

$$\boxed{\delta = 57121}$$

Using this series D. SHANKS computed the value of $\pi$ correct to 707 decimal places.

W. Specht gave an approximation

$$\pi^2 = 9 \cdot 8696\ 044\ldots$$

which yields

$$\text{(41)} \qquad \boxed{\pi \simeq 0 \cdot 26\sqrt{146}} \qquad (5^2 + 11^2 = 146)$$

For a circle with diameter = 1 kilometre, the error for squaring the circle obtained by using (41) is only one millimetre.

We have

(42) $$\boxed{3 \cdot 1415\ 9265\ 358 < \pi < 3 \cdot 1415\ 9265\ 359}$$

and the simple continued fraction expansions of the rational approximations to $\pi$ in (42) are:

(43) $$3.\ 1415\ 9265\ 358 = [3, 7, 15, 1, 292, 1, 1, 1, 1, \ldots]$$

(43′) $$3.\ 1415\ 9265\ 359 = [3, 7, 15, 1, 292, 1, 1, 1, 2, \ldots]$$

Using (42),(43) and (43′) one can find the first *eight* terms of the simple continued fraction for $\pi$:

(44) $$\pi = [3, 7, 15, 1, 292, 1, 1, 1, \ldots]$$

Using the fact that the first eight terms of the continued fractions in (43) and (43') coincide, one can conclude that these eight terms must also occur in the simple continued fraction for $\pi$. Also, the ninth terms can only be = 1, or 2. By closer rational approximations for $\pi$ we can get the further terms of the simple continued fraction expansion for $\pi$.

J. WALLIS, using 35 decimal places, has obtained

(45) $$\boxed{\begin{aligned}\pi = [3, 7, 15, 1, 292, 1, 1, 1, 2, 1, 3, 1, 14, 2, 1, 1,\\ 2, 2, 2, 2, 1, 84, 2, 1, 1, 15, 3, 13,\\ 1, 4, 2, 6, 6, 1, \ldots]\end{aligned}}$$

It is an open problem to discover the law of formation for the coefficients in the continued fractions for $\pi$.

No recognizable pattern has emerged from the expansion (45).

The sequence of convergents to $\pi$ is given by

(46) $$\frac{3}{1}, \frac{22}{7}, \frac{333}{106}, \frac{355}{113}, \frac{103993}{33102}, \frac{104348}{33215}, \ldots$$

The error of approximation can be estimated by means of the inequality (18), [See p.128],

$$|\pi - \frac{22}{7}| < \frac{1}{7 \times 106} < \frac{1}{10^2};$$

$$\left|\pi - \frac{333}{106}\right| < \frac{1}{106 \times 113} < \frac{1}{10^4};$$

$$\left|\pi - \frac{355}{113}\right| < \frac{1}{113 \times 33102} < \frac{1}{10^6};$$

and so on. The sequence of convergents of the continued fraction expansion of a given number provides very good approximations to that number. We have mentioned that the rational approximation

$$(47) \qquad \boxed{\pi \simeq \frac{355}{113}} \qquad \begin{cases} \dfrac{355}{113} & = 3.1415929\ldots \\ \pi & = 3.1415926\ldots \end{cases}$$

is given in Kriyakramakari. In the year 500 A.D. Chao Jung-Tze had obtained the two estimates $\frac{22}{7}, \frac{355}{113}$.

*Remark:* Let us observe that in the rational approximation (47) to $\pi$ each of the first three *odd* numbers 1,3,5 occur *twice*

Srinivasa Ramanujan, [See ≪*Squaring the circle*≫, Jour. Indian Math. See,Vol5, (1913). P.132] gives an interesting geometric construction for finding $\sqrt{\frac{355}{113}}$.

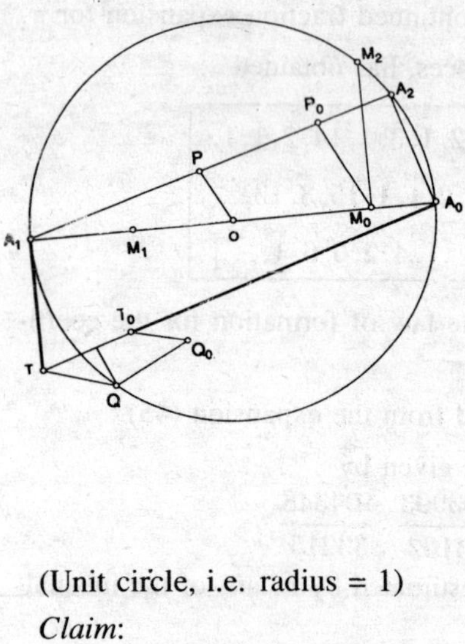


$$M_0M_2 \perp A_1A_0$$

$$A_0A_2 = M_0M_2$$

$$OP \parallel M_0P_0 \parallel A_0A_2$$

$$A_1M_1 = M_10 = \frac{1}{2}$$

$$0M_0 = \frac{2}{3}$$

$A_1T$ is tangent to the circle and

$$A_1T = PP_0$$

$$A_1Q = A_1P$$

$$A_0Q_0 = A_0M_1 = \frac{3}{2}$$

$$Q_0T_0 \parallel QT$$

(Unit circle, i.e. radius = 1)

*Claim*:

$$(48) \qquad A_0T_0 = \sqrt{\frac{355}{113}}$$

Since $M_0M_2^2 = 1 - OM_0^2 = 1 - (\frac{2}{3})^2 = \frac{5}{9}$, we have $M_0M_2 = \frac{\sqrt{5}}{3}$.
Since $A_1A_2^2 = 4 - A_0A_2^2 = 4 - M_0M_2^2 = 4 - \frac{5}{9} = \frac{31}{9}$, we have

$A_1A_2 = \dfrac{\sqrt{31}}{3}$. Since

$$\frac{A_1P}{A_10} = \frac{A_1P_0}{A_1M_0} = \frac{A_1A_2}{A_1A_0},$$

we have

$$\frac{A_1P}{1} = \frac{A_1P_0}{1+\dfrac{2}{3}} = \frac{\dfrac{\sqrt{31}}{3}}{2}$$

so that

$$A_1P = \frac{\sqrt{31}}{6}; A_1P_0 = \frac{5\sqrt{31}}{18}$$

whence follows:

$$PP_0 = A_1P - A_1P_0 = \frac{\sqrt{31}}{18}(5-3) = \frac{\sqrt{31}}{9}$$

Therefore

$$A_0T^2 = 4 + A_1T^2 = 4 + PP_0^2 = 4 + \frac{31}{81} = \frac{355}{81}$$

and

$$A_0Q^2 = 4 - A_1Q^2 = 4 - A_1P^2 = 4 - \frac{31}{36} = \frac{113}{36}$$

so that

$$\frac{A_0T}{A_0Q} = \frac{2}{3}, \sqrt{\frac{355}{113}}$$

Since

$$\frac{A_0T_0}{A_0Q_0} = \frac{A_0T}{A_0Q} \text{ and } A_0Q_0 = A_0M_1 = \frac{3}{2}$$

we get $A_0T_0 = \dfrac{3}{2}\cdot\dfrac{2}{3}\sqrt{\dfrac{355}{113}}$, which proves the claim (48).

## § 7. THEOREM OF BHASKARA

We now proceed to show that Bhaskara's chakravala method of finding the integer solutions of the equation

(1) $$\boxed{Dx^2 + 1 = y^2} \quad 1 < D \neq a \quad \text{square}$$

has an intimate connection with continued fractions.

**Theorem 1. [BHASKARA]**

Let $D$ be a positive integer which is not a square.

(2) $$\boxed{F(x, y) = Dx^2 - y^2}$$

and put

(3) $$(x_0, y_0, h_0) = (1, [\sqrt{D}], F(x_0, y_0))$$

Determine $m_0$ by the conditions

(4) $$m_0 x_0 + y_0 \equiv 0 (\text{mod}\, h_0) \;\; \textit{and} \;\; m_0 < \sqrt{D} < m_0 + h_0$$

Define, inductively, the four sequences of positive integers

$$(m_\lambda, x_\lambda, y_\lambda, h_\lambda) \quad (\lambda = 0, 1, 2, \ldots)$$

by

$$(x_{\lambda+1}, y_{\lambda+1}, h_{\lambda+1}) = \left( \frac{m_\lambda x_\lambda + y_\lambda}{h_\lambda}, \; \frac{D x_\lambda + m_\lambda y_\lambda}{h_\lambda}, \; \frac{D - m_\lambda^2}{h_\lambda} \right)$$

wherein $m_\lambda$ is determined by the conditions:

(6) $$m_\lambda x_\lambda + y_\lambda \equiv 0 (\text{mod}\, h_\lambda) \text{ and } m_\lambda < \sqrt{D} < m_\lambda + h_\lambda$$

Then we claim that:

I° There holds the Bhaskara's identity:

(7) $$\boxed{h_\lambda = (-1)^\lambda F(x_\lambda . y_\lambda)} \quad (\lambda = 0, 1, 2, \ldots)$$

II° There hold the recurrence relations

(8) $$\left| \begin{array}{l} x_\lambda = a_\lambda x_\lambda + x_{\lambda-2} \\ y_\lambda = a_\lambda y_{\lambda-1} + y_{\lambda-2} \end{array} \right| \quad (\lambda \geq 2)$$

wherein the coeffients $a_\lambda$ are given by:

(9) $$\boxed{a_{\lambda+1} = \frac{m_\lambda + m_{\lambda-1}}{h_\lambda}} \quad (\lambda = 1, 2, \ldots)$$

with $(a_0, a_1) = (y_0, x_1)$. There exist only finitely many $a_\lambda$ that are distinct from one another.

III° There exists a value of $\lambda$ such that $h_\lambda = 1$

**Proof: Step 1:** We have

$$F(x_{\lambda+1}, y_{\lambda+1}) = D\left(\frac{m_\lambda x_\lambda + y_\lambda}{h_\lambda}\right)^2 - \left(\frac{D x_\lambda + m_\lambda y_\lambda}{h_\lambda}\right)^2$$

$$= \frac{1}{h_\lambda^2}\left\{ D(m_\lambda^2 x_\lambda^2 + y_\lambda^2 + 2 m_\lambda x_\lambda y_\lambda) - (D^2 x_\lambda^2 + m_\lambda^2 y_\lambda^2 + 2 D x_\lambda m_\lambda y_\lambda \right\}$$

$$= \frac{1}{h_\lambda^2}\left\{ -D x_\lambda^2 (D - m_\lambda^2) + y_\lambda^2 (D - m_\lambda^2) \right\}$$

$$= \frac{1}{h_\lambda^2}(D - m_\lambda^2)(-Dx_\lambda^2 + y_\lambda^2) = -F(x_\lambda y_\lambda)\frac{1}{h_\lambda}h_{\lambda+1}$$

whence follows

$$\frac{F(x_{\lambda+1}, y_{\lambda+1})}{h_{\lambda+1}} = -\frac{F(x_\lambda, y_\lambda)}{h_\lambda}$$

or, what is the same

$$(-1)^\lambda \frac{F(x_\lambda, y_\lambda)}{h_\lambda} = (-1)^{\lambda+1}\frac{F(x_{\lambda-1}, y_{\lambda-1})}{h_{\lambda+1}} \dots$$

whence follows

$$(-1)^\lambda \frac{F(x_\lambda y_\lambda)}{h_\lambda} = (-1)^{\lambda-1}\frac{F(x_{\lambda-1}, y_{\lambda-1})}{h_{\lambda-1}} = \dots = \frac{F(x_0, y_0)}{h_0} = 1,$$

[See (3)]. Consequently $(-1)^\lambda F(x_\lambda, y_\lambda) = h_\lambda$. This proves $I^0$.

(10) **Step 2** | We prove that $x_\lambda, y_\lambda, h_\lambda$ are always mutually coprime integers. $(\lambda = 0, 1, 2, \dots)$

For $\lambda = 0$, we have $(x_0, y_0, h_0) = (1, [\sqrt{D}], Dx_0^2 - y_0^2)$, thus the assertion holds for $\lambda = 0$. We prove it by induction on $\lambda$. Assuming (10) for $\lambda$, the condition (6) shows at once that $x_{\lambda+1}$ is an integer. We want to show that $y_{\lambda+1}, h_{\lambda+1}$ are also integers. Since, by (6),

$$y_\lambda \equiv -m_\lambda x_\lambda (\text{mod}\, h_\lambda)$$

and since $Dx_\lambda^2 - y_\lambda^2 = (-1)^\lambda h_\lambda$ [See (6), just proved above]

$$Dx_\lambda^2 \equiv y_\lambda^2 (\text{mod}\, h_\lambda)$$

so that

$$Dx_\lambda^2 \equiv y_\lambda^2 \equiv (-m_\lambda x_\lambda)^2 \equiv m_\lambda^2 x_\lambda^2 (\text{mod}\, h_\lambda)$$

Since $x_\lambda$ is coprime with $h_\lambda$ (by induction assumption) there follows

$$D \equiv m_\lambda^2 (\text{mod}\, h_\lambda)$$

i.e. $\dfrac{D - m_\lambda^2}{h_\lambda} = h_{\lambda+1}$ is an integer. Next

$$Dx_\lambda + m_\lambda y_\lambda \equiv m_\lambda^2 . x_\lambda + m_\lambda(-m_\lambda x_\lambda) \equiv 0 (\text{mod}\, h_\lambda)$$

i.e. $\dfrac{Dx_\lambda + m_\lambda y_\lambda}{h_\lambda} = y_{\lambda+1}$ is an integer.

Since

$$x_\lambda y_{\lambda+1} - x_{\lambda+1} y_\lambda = x_\lambda \frac{Dx_\lambda + m_\lambda y_\lambda}{h_\lambda} - \frac{m_\lambda x_\lambda + y_\lambda}{h_\lambda} . y_\lambda$$

$$= \frac{1}{h_\lambda}(Dx_\lambda^2 - y_\lambda^2) = (-1)^\lambda$$

we get the relation:

(11) $$x_\lambda y_{\lambda+1} - x_{\lambda+1} y_\lambda = (-1)^\lambda$$

which shows that $x_{\lambda+1}, y_{\lambda+1}$, must be coprime. then the relation $Dx^2_{\lambda+1} - y^2_{\lambda+1} = (-1)^{\lambda+1} h_{\lambda+1}$ shows that $h_{\lambda+1}$ is coprime with $x_{\lambda+1}, y_{\lambda+1}$. The induction is complete and (10) is thereby proved.

**Step 3**: From (3) and (9) we have: $(x_0, y_0) = (1, a_0)$, therefore we have $h_0 = Dx_0^2 - y_0^2 = D.1^2 - a_0^2 = D - a_0^2$, thus

$$\sqrt{D} - a_0 = \frac{h_0}{\sqrt{D} + a_0}$$

which we can write as

(12) $$\sqrt{D} = a_0 + \frac{1}{\delta_1} \quad \text{with} \quad \delta_1 = \frac{\sqrt{D} + a_0}{h_0}$$

Since $a_0 < \sqrt{D} < a_0 + 1$, thus $0 < \sqrt{D} - a_0 < 1$, we see that $\delta_1 > 1$

Now $D - m_0^2 = h_0 h_1$ so that

$$\frac{\sqrt{D} + m_0}{h_1} = \frac{h_0}{\sqrt{D} - m_0} =: \delta_2 \text{ (say)}$$

then

$$\delta_1 - \frac{1}{\delta_2} = \frac{\sqrt{D} + a_0}{h_0} - \frac{\sqrt{D} - m_0}{h_0} = \frac{a_0 + m_0}{h_0} =: a_1$$

Hence we have

(13)

$$\delta_1 = a_1 + \frac{1}{\delta_2} \quad \text{with} \quad \delta_2 = \frac{\sqrt{D} + m_0}{h_1} \quad \text{and} \quad a_1 = \frac{m_0 + a_0}{h_0}$$

Since $m_0 < \sqrt{D} < m_0 + h_0$ we have $0 < \dfrac{\sqrt{D} - m_0}{h_0} < 1, 0 < \dfrac{1}{\delta_2} < 1$, thus $\delta_2 > 1$

Let us set

(14) $$\delta_\lambda = \frac{\sqrt{D} + m_{\lambda-2}}{h_{\lambda-1}} \quad (\lambda = 2, 3, \ldots)$$

Then it follows from the relation $D - m^2_{\lambda-1} = h_{\lambda-1} h_\lambda$ that

$$\frac{h_\lambda}{\sqrt{D} + m_{\lambda-1}} = \frac{\sqrt{D} - m_{\lambda-1}}{h_{\lambda-1}}, \quad \text{i.e.} \quad \frac{1}{\delta_{\lambda+1}} = \frac{\sqrt{D} - m_{\lambda-1}}{h_{\lambda-1}}$$

and

$$\delta_\lambda - \frac{1}{\delta_{\lambda+1}} = \frac{m_{\lambda-1} + m_{\lambda-2}}{h_{\lambda-1}} = a_\lambda, \text{ [See (9)]}$$

thus

(15) $$\boxed{\delta_\lambda = a_\lambda + \frac{1}{\delta_{\lambda+1}}}$$

and

(16) $$\boxed{a_\lambda = \frac{m_{\lambda-1} + m_{\lambda-2}}{h_{\lambda-1}}} \qquad (\lambda = 2, 3, \ldots)$$

Since $m_{\lambda-1}$ was chosen with $m_{\lambda-1} < \sqrt{D} < m_{\lambda-1} + h_{\lambda-1}$, thus $0 < \dfrac{\sqrt{D} - m_{\lambda-1}}{h_{\lambda-1}} < 1$, i.e $0 < \dfrac{1}{\delta_{\lambda+1}} < 1$ $(\lambda = 1, 2, \ldots)$ we always have

(17) $$\delta_\lambda > 1 \qquad (\lambda = 1, 2, \ldots)$$

since we already have seen above in (12) that $\delta_1 > 1$.

**Step 4** | We prove that the numbers $a_\lambda (\lambda = 0, 1, 2, \ldots)$ are all *integers*.

Since $(x_0, y_0) = (1, [\sqrt{D}])$, and $a_0 = y_0$, $a_0$ is an integer.

Since $x_1 = \dfrac{m_0 x_0 + y_0}{h_0}$ is an integer, and $a_1 = \dfrac{a_0 + m_0}{h_0} = x_1$ it follows that $a_1$ is an integer. In order to show that $a_\lambda$, defined by (16), are integers we have to show:

(18) $$\boxed{m_\lambda + m_{\lambda-1} \equiv 0 (\operatorname{mod} h_\lambda)} \qquad (\lambda = 1, 2, \ldots)$$

**Proof of (18)** We have, since $D - m_{\lambda-1}^2 = h_\lambda h_{\lambda+1}$,

$$\begin{aligned} y_\lambda &= \frac{D x_{\lambda-1} + m_{\lambda-1} y_{\lambda-1}}{h_{\lambda-1}} \\ &= \frac{1}{h_{\lambda-1}} \left[ (m_{\lambda-1}^2 + h_{\lambda-1} h_\lambda) x_{\lambda-1} + m_{\lambda-1} y_{\lambda-1} \right] \\ &= \frac{1}{h_{\lambda-1}} \left[ m_{\lambda-1} (m_{\lambda-1} x_{\lambda-1} + y_{\lambda-1}) + h_{\lambda-1} h_\lambda x_{\lambda-1} \right] \\ &= m_{\lambda-1} \frac{m_{\lambda-1} x_{\lambda-1} + y_{\lambda-1}}{h_{\lambda-1}} + h_\lambda x_{\lambda-1} = m_{\lambda-1} x_\lambda + h_\lambda x_{\lambda-1} \end{aligned}$$

which gives

$$y_\lambda = m_{\lambda-1}x_\lambda + h_\lambda x_{\lambda-1} \text{ so that } m_{\lambda-1}x_\lambda \equiv y_\lambda (\text{mod}\, h_\lambda)$$

Since always, $m_\lambda x_\lambda + y_\lambda \equiv 0(\text{mod}\, h_\lambda)$ we have

$$m_\lambda x_\lambda \equiv -y_\lambda (\text{mod}\, h_\lambda)$$

which, together with (19), yields $(m_{\lambda-1} + m_\lambda)x_\lambda \equiv 0(\text{mod}\, h_\lambda)$. Since $(x_\lambda, h_\lambda) = 1$, we obtain (18).

From (12), (13), and (15) we therefore have

(20)
$$\left|\begin{aligned} \sqrt{D} &= a_0 + \frac{1}{\delta_1} \\ \delta_1 &= a_1 + \frac{1}{\delta_2} \\ &\vdots \\ \delta_\lambda &= a_\lambda + \frac{1}{\delta_{\lambda+1}} \\ &\vdots \end{aligned}\right. \qquad \begin{aligned} &a_\lambda \text{ are all } \textit{integers} \\ &\delta_\lambda > 1 (\lambda = 1, 2, \ldots) \end{aligned}$$

In other words:

> $a_0, a_1, \ldots a_\lambda, \ldots$ are the coefficients of the continued fraction expansion of $\sqrt{D}$, and $\delta_\lambda$ are its complete quotients, thus

(21) $$a_\lambda = [\delta_\lambda]$$

**Step 5:** Since, always, $0 < m_\lambda < \sqrt{D}$, we have from (16), (since $h_\lambda$ are positive integers).

(22) $$\boxed{0 < a_\lambda < 2\sqrt{D}}$$

Since the number of *positive* integers beneath any prescribed bound is necessarily *finite* it follows from (22) that there exist only finitely many $a_\lambda$ that are distinct from one another.

Again, since,

$$\delta_{\lambda+1} = \frac{\sqrt{D} + m_{r-1}}{h_\lambda} > 1 \text{ and } 0 < m_{\lambda-1} < \sqrt{D}$$

we find

$$h_\lambda < \sqrt{D} + m_{\lambda-1} < 2\sqrt{D}$$

thus

(22) $$\boxed{0 < h_\lambda < 2\sqrt{D}}$$

Since $m_\lambda$ and $h_\lambda$ are positive integers, and

$$0 < m_\lambda < \sqrt{D} \text{ and } 0 < h_\lambda < 2\sqrt{D}$$

it follows that only finitely many among the $m_\lambda$ (resp: $h_\lambda$) can be distinct from one another.

Consequently,

(24) there exist only finitely many distinct complete quotients $\delta_\lambda (\lambda = 1, 2, \ldots)$

In the sequence of complete quotients, let p be the *first* index such that $\delta_p$ repeats itself:

$$\delta_1, \delta_2, \ldots, \delta_p, \ldots, \delta_{p+N} \cdots$$

$$\delta_p = \delta_{p+N} \tag{25}$$

Since

$$\delta_p = a_p + \frac{1}{\delta_{p+1}}, \delta_{p+N} = \delta_{p+N} + \frac{1}{\delta_{p+N+1}}, \tag{26}$$

and

$$a_p = [\delta_p] = [\delta_{p+N}] = a_{p+N}$$

it follows from (26) that $\delta_{p+1} = \delta_{p+n+1}$, whence

$$a_{p+1} = a_{p+N+1},$$

since these are the integral parts of $\delta_{p+1} = \delta_{p+n+1}$.
Similarly for $\delta_{p+2}, \delta_{p+3}, \ldots, i.e.$

$$\left.\begin{aligned} \delta_{p+N+k} &= \delta_{p+k} \\ a_{p+N+k} &= a_{p+k} \end{aligned}\right\} \quad (k = 1, 2, \ldots) \tag{27}$$

in other words

(28) The continued fraction expansion of $\sqrt{D}$ is ultimately *periodic*, i.e.
$\sqrt{D} = [a_0, a_1, \ldots a_{p-1}, \overset{*}{a}_p \ldots, \overset{*}{a}_{p+N}]$

**Step 6**: We show that in (28) one must have p=1. In the argument below we are going to make use of the following:

*Remark*: Given a quadratic surd $\theta = a + b\sqrt{D}, D > 0, (D \neq$ a square integer), denote by $\overline{\theta}$ its conjugate and put $\theta' = -\frac{1}{\overline{\theta}}$

$$\overline{\theta} = a - b\sqrt{D}, \theta' = -\frac{1}{\overline{\theta}} = \frac{1}{-a + b\sqrt{D}}$$

Since $a_1 + b_1\sqrt{D} = a_2 + b_2\sqrt{D}$ if and only $a_1 = a_2$, and $b_1 = b_2$ we always have

$$\theta_1 = \theta_2 \Rightarrow \overline{\theta}_1 = \overline{\theta}_2 \Rightarrow \theta'_1 = \theta'_2$$

Since $\delta_\lambda = a_\lambda + \dfrac{1}{\delta_{\lambda+1}}$, we get in virtue of the remark,

$$\overline{\delta}_\lambda = a_\lambda + \frac{1}{\overline{\delta}_{\lambda+1}} \text{ or } -\frac{1}{\delta'_\lambda} = a_\lambda - \delta'_{\lambda+1}$$

which can be written

(29) $$\delta'_{\lambda+1} = a_\lambda + \frac{1}{\delta'_\lambda}$$

Now

$$\delta_{\lambda+1} = \frac{\sqrt{D} + m_{\lambda-1}}{h_\lambda}, \overline{\delta}_{\lambda+1} = \frac{-\sqrt{D} + m_{\lambda-1}}{h_\lambda}, \delta'_{\lambda+1} = \frac{h_\lambda}{\sqrt{D} - m_{\lambda-1}}$$

Since always. $0 < m_\lambda < \sqrt{D}$, we have $\delta'_{\lambda+1} > 0$; changing $\lambda$ to $\lambda - 1$ we must also have $\delta'_\lambda > 0$. Then from (29) follows:

$$\delta'_{\lambda+1} > a_\lambda \geq 1$$

Since $\lambda$ is arbitrary, we thus conclude:

$$\delta'_\lambda > 1 \text{ for all } \lambda$$

Consequently, from (29), $a_\lambda$ turns out to be the integral part of $\delta'_{\lambda+1}$

(30) $$\boxed{[\delta_\lambda] = a_\lambda = [\delta'_{\lambda+1}]}$$

Now $p$ was defined as the *first* index such that

(31) $$\delta_{p+N} = \delta_p \quad \text{(See (25))}$$

But then

$$\delta'_{p+N} = \delta'_p \text{ so that } [\delta'_{p+N}] = [\delta'_p]$$

which yields, because of (30)

(32) $$a_{p+N-1} = a_{P-1}$$

From (31) and (32)

(32) $$a_{p+N-1} + \frac{1}{\delta_{p+N}} = a_{p-1} + \frac{1}{\delta_p}$$

or, what is the same,

$$\delta_{(p-1)+N} = \delta_{p-1}$$

contradicting that $p \geq 1$ was the least index with the property (31). Consequently $p = 1$, and the continued fraction expansion of $\sqrt{D}$ must be of the form

(33) $$\boxed{\sqrt{D} = [a_0; \overset{*}{a}_1, a_2, \ldots, \overset{*}{a}_N]}$$

**Proof of I° in Theorem 1.**

We solve for $x_\lambda, y_\lambda$ from

$$m_\lambda x_\lambda + y_\lambda = h_\lambda x_{\lambda+1}$$
$$Dx_\lambda + m_\lambda y_\lambda = h_\lambda y_{\lambda+1}$$

and obtain

(34) $$x_\lambda = \frac{y_{\lambda+1} - m_\lambda x_{\lambda+1}}{h_{\lambda+1}}, \quad y_\lambda = \frac{Dx_{\lambda+1} - m_\lambda y_{\lambda+1}}{h_{\lambda+1}}$$

Since, by definition,

$$x_\lambda = \frac{m_{\lambda-1}x_{\lambda-1} + y_{\lambda-1}}{h_{\lambda-1}} \text{ and } a_\lambda = \frac{m_{\lambda-1} + m_{\lambda-2}}{h_{\lambda-1}}$$

We can write

$$\begin{aligned} x_\lambda &= \left(a_\lambda - \frac{m_{\lambda-2}}{h_{\lambda-1}}\right) \cdot x_{\lambda-1} + \frac{y_{\lambda-1}}{h_{\lambda-1}} \\ &= a_\lambda x_{\lambda-1} + \frac{y_{\lambda-1} - m_{\lambda-2}x_{\lambda-1}}{h_{\lambda-1}} \\ &= a_\lambda x_{\lambda-1} + x_{\lambda-2} \quad \text{[See (34)]} \end{aligned}$$

Again

$$\begin{aligned} y_\lambda &= \frac{Dx_{\lambda-1} + m_{\lambda-1}y_{\lambda-1}}{h_{\lambda-1}} = \frac{Dx_{\lambda-1}}{h_{\lambda-1}} + (a_\lambda - \frac{m_{\lambda-2}}{h_{\lambda-1}})y_{\lambda-1} \\ &= a_\lambda y_{\lambda-1} + \frac{Dx_{\lambda-1} - m_{\lambda-2}y_{\lambda-1}}{h_{\lambda-1}} \\ &= a_\lambda y_{\lambda-1} + y_{\lambda-2} \quad \text{[See (34)]} \end{aligned}$$

Accordingly the $x_\lambda$ and $y_\lambda$ satisfy the same recurrence relations (8) as the $q_\lambda$ and $p_\lambda$. We verify that they have the same initial values for $\lambda = 0, 1$.

Firstly,

(35) $$(x_0, y_0) = (1, [\sqrt{D}]) = (q_0 . p_0) \text{ and } a_0 = y_0$$

and

$$x_1 = \frac{m_0 x_0 + y_0}{h_0} = \frac{m_0 + a_0}{h_0} = a_1 \quad \text{(See (13))}$$

By (11), $x_0 y_1 = x_1 y_0 = 1$, thus $1.y_1 - a_1 a_0 = 1$, i.e.

(36) $$(x_1, y_1) = (a_1, a_1 a_0 + 1) = (q_1, p_1)$$

It follows therefore

$$(x_\lambda, y_\lambda) = (q_\lambda, p_\lambda) \quad \lambda = 0, 1, 2, \ldots \tag{37}$$

**Proof of III° in Theorem 1.**

Since we have $\delta_1 = \delta_{N+1}, \sqrt{D} = [a_0, a_1, \ldots, a_n, \delta_{N+1}]$, we get

$$\sqrt{D} = \frac{\delta_{N+1}p_N + p_{N-1}}{\delta_{N+1}q_N + q_{N-1}} = \frac{\delta_1 p_N + p_{N-1}}{\delta_1 q_N + q_{N-1}} = \frac{p_N + \dfrac{1}{\delta_1}p_{N-1}}{q_N + \dfrac{1}{\delta_1}q_{N-1}}$$

Since $\sqrt{D} = a_0 + \dfrac{1}{\delta_1}$ this gives

$$\sqrt{D}[q_N + (\sqrt{D} - a_0)q_{N-1}] = p_N + (\sqrt{D} - a_0)p_{N-1}$$

Comparing the rational and irrational parts on both sides, we have

$$\begin{aligned} Dq_{N-1} &= p_N - a_0 p_{N-1} \\ p_{N-1} &= q_N - a_0 q_{N-1} \end{aligned} \tag{38}$$

Multiplying the first equation by $q_{N-1}$, the second by $p_{N-1}$ and subtracting we get

$$Dq_{N-1}^2 - p_{N-1}^2 = p_N q_{N-1} - p_{N-1} q_N = (-1)^{N-9}$$

since $(x_{N-1}, y_{N-1}) = (q_{N-1}, p_{N-1}$, we get

$$(-1)^{N-1}[Dx_{N-1}^2 - y_{N-1}^2] = 1$$

i.e.,

$$\boxed{h_{N-1} = 1}$$

All the three claims I°, II° and III° of Bhaskara's theorem are established, and the proof is hereby completed. ■

**Scholium**

We shall summarize here the main formulas establishing the connection between Bhaskara's Chakravala method for solving the equation $Dx^2 + 1 = y^2, D > 1, D \neq a$ square, and the continued fraction expansion of $\sqrt{D}$.

$$
\begin{aligned}
\sqrt{D} &= a_0 + \frac{1}{\delta_1}, & \delta_1 &= \frac{\sqrt{D}+a_0}{h_0} \\
\delta_1 &= a_1 + \frac{1}{\delta_2}, & \delta_2 &= \frac{\sqrt{D}+m_0}{h_1} \\
&\vdots\ \vdots & &\vdots \\
\delta_\lambda &= a_\lambda + \frac{1}{\delta_{\lambda+1}}, & \boxed{\delta_\lambda = \frac{\sqrt{D}+m_{\lambda-2}}{h_{\lambda-1}}} & \quad (\lambda = 2, 3, \ldots) \\
&\vdots\ \vdots\ \vdots & &\vdots
\end{aligned} \tag{39}
$$

$$
a_0 = [\sqrt{D}], \quad a_1 = \frac{m_0 + a_0}{h_0}, \quad \boxed{a_\lambda = \frac{m_{\lambda-1}+m_{\lambda-2}}{h_{\lambda-1}}} \tag{40}
$$

$$
(x_{\lambda+1}, y_{\lambda+1}, h_{\lambda+1}) = \left(\frac{m_\lambda x_\lambda + y_\lambda}{h_\lambda}, \frac{Dx_\lambda + m_\lambda y_\lambda}{h_\lambda}, \frac{D - m_\lambda^2}{h_\lambda}\right)
$$

$$
(x_0, y_0) = (1, [\sqrt{D}])
$$

$$
\boxed{\begin{aligned} m_\lambda x_\lambda + y_\lambda \equiv 0 (\text{mod } h_\lambda), m_\lambda < \sqrt{D} < m_\lambda + h_\lambda \\ h_\lambda = (-1)^\lambda [Dx_\lambda^2 - y_\lambda^2] \end{aligned}} \tag{41}
$$

The Bhaskara multipliers $m_\lambda$ occur in the numerators of the complete quo.ients $\delta_\lambda$; and they are connected with the coefficients $a_\lambda$ of the continued fraction for $\sqrt{D}$ by (40). the $h_\lambda$ occur in the denominators of $\delta_\lambda$.

$$
\boxed{\begin{array}{l|l} (x_\lambda, y_\lambda) = (q_\lambda, p_\lambda) & p_\lambda = a_\lambda p_{\lambda-1} + p_{\lambda-2} \\ (x_0, y_0) = (1, \sqrt{D}) = (1, a_0) & q_x = a_\lambda q_{\lambda-1} + q_{\lambda-2} \\ & \qquad (\lambda = 2, 3, \ldots) \end{array}} \tag{42}
$$

**Example 1.** $\underline{D = 67}$, $a_0 = [\sqrt{D}] = 8$

$$
\sqrt{D} = \sqrt{67} = 8 + (\sqrt{67} - 8) = 8 + \frac{3}{\sqrt{67}+8}
$$

$$
\delta_1 = \frac{\sqrt{67}+8}{3} = 5 + \frac{\sqrt{67}-7}{3} = 5 + \frac{6}{\sqrt{67}+7}
$$

$$
\delta_2 = \frac{\sqrt{67}+7}{6} = 2 + \frac{\sqrt{67}-5}{6} = 2 + \frac{7}{\sqrt{67}+5}
$$

$$\delta_3 = \frac{\sqrt{67}+5}{7} = 1 + \frac{\sqrt{67}-2}{7} = 1 + \frac{9}{\sqrt{67}+2}$$

$$\delta_4 = \frac{\sqrt{67}+2}{9} = 1 + \frac{\sqrt{67}-7}{9} = 1 + \frac{2}{\sqrt{67}+7}$$

(43) $$\delta_5 = \frac{\sqrt{67}+7}{2} = 7 + \frac{\sqrt{67}-7}{2} = 7 + \frac{9}{\sqrt{67}+7}$$

$$\delta_6 = \frac{\sqrt{67}+7}{9} = 1 + \frac{\sqrt{67}-2}{9} = 1 + \frac{7}{\sqrt{67}+2}$$

$$\delta_7 = \frac{\sqrt{67}+2}{7} = 1 + \frac{\sqrt{67}-5}{7} = 1 + \frac{6}{\sqrt{67}+5}$$

$$\delta_8 = \frac{\sqrt{67}+5}{6} = 2 + \frac{\sqrt{67}-7}{6} = 2 + \frac{3}{\sqrt{67}+7}$$

$$\delta_9 = \frac{\sqrt{67}+7}{3} = 5 + \frac{\sqrt{67}-8}{3} = 5 + \frac{1}{\sqrt{67}+8}$$

$$\delta_{10} = \frac{\sqrt{67}+8}{1} = 16 + (\sqrt{67}-8) = 16 + \frac{3}{\sqrt{67}+8}$$

$$\delta_{11} = \frac{\sqrt{67}+8}{3} = \delta_1$$

(44)

| $\lambda$ | 0 | 1 | 2 | 3 | 4 | 5 | 6 | 7 | 8 | 9 | 10 |
|---|---|---|---|---|---|---|---|---|---|---|---|
| $m_\lambda$ | 7 | 5 | 2 | 7 | 7 | 2 | 5 | 7 | 8 | 8 | 7 |
| $x_\lambda$ | 1 | 5 | 11 | 16 | 27 | 205 | 232 | 437 | 1106 | **5967** | 96578 |
| $y_\lambda$ | 8 | 41 | 90 | 131 | 221 | 1678 | 1899 | 3577 | 9053 | **48842** | 790525 |
| $h_\lambda$ | 3 | 6 | 7 | 9 | 2 | 9 | 7 | 6 | 3 | **1** | 3 |
| $a_\lambda$ | 8 | 5 | 2 | 1 | 1 | 7 | 1 | 1 | 2 | 5 | 16 |

(44′) $$\sqrt{67} = [8; \overset{*}{5}, 2, 1, 1, 7, 1, 1, 2, 5, \overset{*}{16}]$$

$h_9 = 1$, i.e. $(-1)^9[Dx_9^2 - y_9^2] = 1$, or $Dx_9^2 + 1 = y_9^2$

$$67 \cdot (\mathbf{5967})^2 + 1 = (\mathbf{48842})^2 \quad \text{[See (19), p. 114]}$$

We compute the continued fraction expansions of $\sqrt{D}$,

$$D = 61, 103, 97$$

as in (43), (44′) in Example 1. We shall not write down the details, but display the obtained results in a tabular form as in (44).

*Example 2.* $\underline{D = 61,}$ $a_0 = [\sqrt{D}] = 7$

(45)

| $\lambda$ | 0 | 1 | 2 | 3 | 4 | 5 | 6 | 7 | 8 | 9 | 10 | 11 |
|---|---|---|---|---|---|---|---|---|---|---|---|---|
| $m_\lambda$ | 5 | 7 | 5 | 4 | 6 | 4 | 5 | 7 | 5 | 7 | 7 | 7 |
| $x_\lambda$ | 1 | 1 | 5 | 16 | 21 | 58 | 137 | 195 | 722 | 3083 | **3805** | 56353 |
| $y_\lambda$ | 7 | 8 | 39 | 125 | 164 | 453 | 1070 | 1523 | 5639 | 24079 | **29718** | 440131 |
| $h_\lambda$ | 12 | 3 | 4 | 9 | 5 | 5 | 9 | 4 | 3 | 12 | **1** | 12 |
| $a_\lambda$ | 7 | 1 | 4 | 3 | 1 | 2 | 2 | 1 | 3 | 4 | 1 | 14 |

(45) $$\sqrt{61} = [7; \overset{*}{1}, 4, 3, 1, 2, 2, 1, 3, 4, 1, \overset{*}{14}]$$

Since $h_{10} = 1, Dx_{10}^2 - y_{10}^2 = 1, Dx_{10}^2 - 1 = y_{10}^2$

$$61 \cdot (\mathbf{3805})^2 - 1 = (\mathbf{29718})^2, See (22'), P. 113$$

*Example 3:* $\underline{D = 97,}$ $a_0 = [\sqrt{D}] = 9$

(46)

| $\lambda$ | 0 | 1 | 2 | 3 | 4 | 5 | 6 | 7 | 8 | 9 | 10 | 11 |
|---|---|---|---|---|---|---|---|---|---|---|---|---|
| $m_\lambda$ | 7 | 8 | 3 | 5 | 4 | 5 | 3 | 8 | 7 | 9 | 9 | 7 |
| $x_\lambda$ | 1 | 1 | 6 | 7 | 13 | 20 | 33 | 53 | 86 | 483 | **569** | 10725 |
| $y_\lambda$ | 9 | 10 | 59 | 69 | 128 | 197 | 325 | 522 | 847 | 4757 | **5604** | 105629 |
| $h_\lambda$ | 16 | 3 | 11 | 8 | 9 | 9 | 8 | 11 | 3 | 16 | **1** | 16 |
| $a_\lambda$ | 9 | 1 | 5 | 1 | 1 | 1 | 1 | 1 | 1 | 5 | 1 | 18 |

(46′) $$\sqrt{97} = [9; \overset{*}{1}, 5, 1, 1, 1, 1, 1, 1, 5, 1, \overset{*}{18}]$$

Since $h_{10} = 1, Dx_{10}^2 - y_{10}^2 = 1, Dx_{10}^2 - 1 = y_{10}^2$

$$97 \cdot (\mathbf{569})^2 - 1 = (\mathbf{5604})^2, \quad [See P. 117]$$

*Example 4:* $\underline{D = 103,}$ $a_0 = [\sqrt{D}] = 10$

(47)

| $\lambda$ | 0 | 1 | 2 | 3 | 4 | 5 | 6 | 7 | 8 | 9 | 10 | 11 | 12 |
|---|---|---|---|---|---|---|---|---|---|---|---|---|---|
| $m_\lambda$ | 8 | 5 | 7 | 2 | 9 | 9 | 2 | 7 | 5 | 8 | 10 | 10 | 8 |
| $x_\lambda$ | 1 | 6 | 7 | 20 | 27 | 47 | 450 | 497 | 947 | 2391 | 3338 | **22419** | 451718 |
| $y_\lambda$ | 10 | 61 | 71 | 203 | 274 | 477 | 4567 | 5044 | 9611 | 24266 | 33877 | **227528** | 4584437 |
| $h_\lambda$ | 3 | 13 | 6 | 9 | 11 | 2 | 11 | 9 | 6 | 13 | 3 | **1** | 3 |
| $a_\lambda$ | 10 | 6 | 1 | 2 | 1 | 1 | 9 | 1 | 1 | 2 | 1 | 6 | 20 |

(47′) $$\sqrt{103} = [10; \overset{*}{6}, 1, 2, 1, 1, 9, 1, 1, 2, 1, 6, \overset{*}{20}]$$

Since $h_{11} = 1, (-1)^1 1[Dx_{11}^2 - y_{11}^2] = 1, Dx_{11}^2 + 1 = Y_{11}^2$

$$103 \cdot (\mathbf{22419})^2 + 1 = (\mathbf{227528})^2, \quad [\textit{See } (24), P.\ 116]$$

*Remark:* In the continued fraction expansions (44′), (45′), (46′) and (47′) we notice a property of symmetry: namely, if $D > 0$ is an integer $\neq$ a square, $\sqrt{D}$ has a continued fraction expansion of the form

$$\boxed{\sqrt{D} = [a_0; \overset{*}{a_1}, a_2, a_3, \cdots, a_3, a_2, a_1, 2\overset{*}{a_0}]}$$

i.e. the periodic part $\{\overset{*}{a_1}, a_2, \cdots, a_{N-1}, 2\overset{*}{a_0}\}$ has the property of symmetry.

$$a_N = 2a_0, a_k = a_{N-k}, \quad (k = 1, 2, \ldots, N-1)$$

It is true in general, but we shall not prove this fact here.

CHAPTER IV

# SELECTED TOPICS IN GEOMETRY

In this Chapter we prove some propositions in the Geometry of the triangle, the quadrilateral and the circle. It is not our intention here to trace the evolution of geometry in its historical development in the country. For this purpose the reader is referred to the masterly treatment of the topic in the classic treatise of Datta and Singh [3].

## § 1. GEOMETRY IN THE ŚULBA SUTRAS

The root word ≪SULB≫ or ≪ŚULV≫ means: *to measure*, and SULBA means *measuring rope or cord*. However, it is the word ≪RAJJU≫, and not ≪ŚULBA≫ that occurs in the Sutras themselves e.g.,

रज्जुसमासं वक्ष्यामः (कात्यायन शुल्बसूत्र)

(Manipulation with chords will now be expounded)

The Kātyāyana Śulbasutras belong to the Śukla Yajurveda, and the remaining Ŝulbasutras to Kṛṣṇa Yajurveda. The meticulous prescriptions for the constructions of the Védis reveal a precise knowledge of several theorems of elementary Geometry.

The date of the oldest Śulbasutras, the Bodhayana is said to be 8th century B.C.

The theorem stating that the square on the hypotenuse of a right angled triangle is equal to the sum of the squares on its sides has been explicitely stated in the Śulbasutras: it is usually attributed to Pythagoras (c.540 B.C.). We call it the ≪*HYPOTENUSE THEOREM*≫. Many different proofs have been found for the *Hypotenuse Theorem*: We shall give here three proofs of:

**Theorem 1. [The Hypotenuse Theorem]**

If $ABC$ is a right triangle wlith $\hat{C} = \frac{\pi}{2}$ and sides $BC = a$ and $CA = b$, then the hypotenuse $AB = c$ is given by

$$\boxed{c^2 = a^2 + b^2}$$

**Proof [EUCLID]**

We have: $BC = a, CA = b, AB = c$

$CP \perp AB$

$AP = x, BP = y, c = x + y$

$AM = b, AM \perp AC$

$AA' = c, AA' \perp AB$

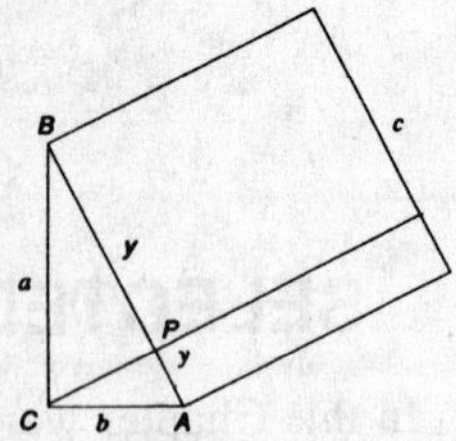


$$cx = b^2, cy = a^2$$

In the two triangles $MAB$ and $CAA'$

$MA = b = CA$

$AB = c = AA'$

$\widehat{MAB} = \frac{\pi}{2} + \hat{A} = \widehat{CAA'}$

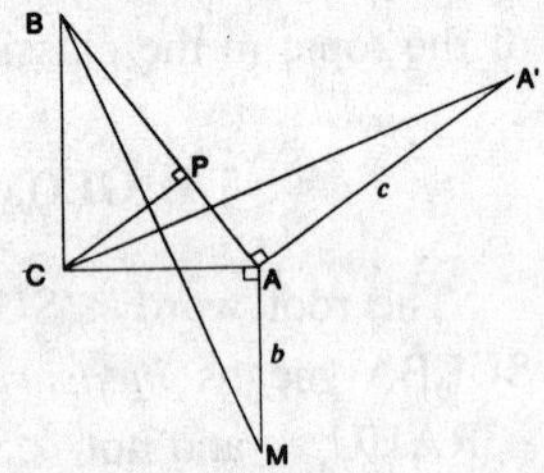


Hence the two triangles are congruent, in particular they have the same area:

$MAB \equiv CAA'$

$$\text{Area } (MAB) = \frac{1}{2} \text{ base } MA.\{ \text{ the distance of } B \text{ from } MA \}$$

$$= \frac{1}{2} \cdot b\{ \text{ the distance of } C \text{ from } MA \},$$

since $BC \parallel AM$

$$= \frac{1}{2} b \cdot b = \frac{1}{2} b^2,$$

and

$$\text{Area } (CAA') = \frac{1}{2} \text{ base } AA'.\{ \text{ the distance of } C \text{ from } AA' \}$$

$$= \frac{1}{2} c \cdot \{ \text{ distance of } P \text{ from } AA' \}$$

$$= \frac{1}{2} c \cdot AP = \frac{1}{2} c \cdot x$$

so that we obtain $\frac{1}{2} cx = \frac{1}{2} b^2, cx = b^2$ :

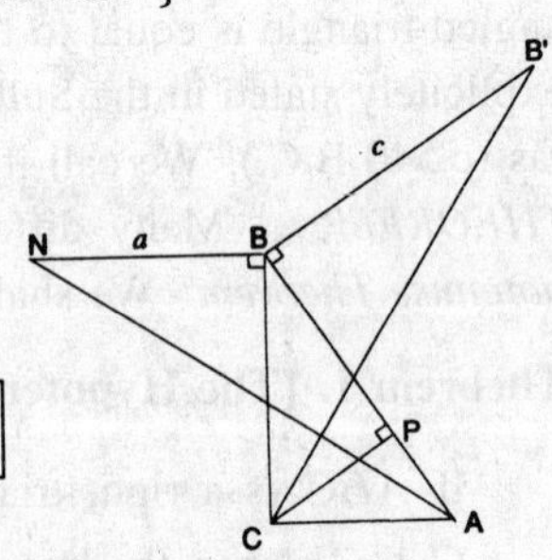


(1) $$cx = b^2,$$

Similarly, in the triangles $NAB$ and $CB'B$ we have

$NB = a = CB$

$$AB = c = B'B$$

$$N\hat{A}B = \frac{\pi}{2} + \hat{B} = C\hat{B}B'$$

Hence $NAB \equiv CB'B$, in particular, Area $(NAB)$ = Area $(CBB')$

Area $(NAB) = \frac{1}{2}$ base $NB.\{$ the distance of $A$ from $NB$ $\}$

$= \frac{1}{2}a \cdot \{$ the distance of $C$ from $NB$ $\}$,
since $NB \parallel CA$

$= \frac{1}{2}a \cdot a = \frac{1}{2}a^2,$

Area $(CB'B) = \frac{1}{2}$ base $BB'\{$ the distance of $C$ from $BB'$ $\}$

$= \frac{1}{2}c\{$ the distance of $P$ from $BB'$ $\}$

$= \frac{1}{2}c \cdot BP = \frac{1}{2}c \cdot y$

so that we obtain $\frac{1}{2}cy = \frac{1}{2}a^2, cy = a^2$

(2) $\boxed{cy = a^2}$

From (1) and (2):

$$c(x+y) = a^2 + b^2, \text{ i.e. } c^2 = a^2 + b^2$$

*Remark on Euclid's proof:*

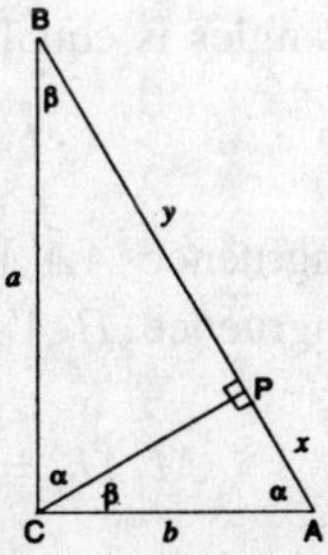


We find that the triangles $ABC$ and $ACP$ are similar, as also the triangles $ABC$ and $CBP$; since $ABC \parallel\!| ACP$,

$$\frac{AB}{AC} = \frac{AC}{AP}, \quad \frac{c}{b} = \frac{b}{x}, \quad \underline{cx = b^2}$$

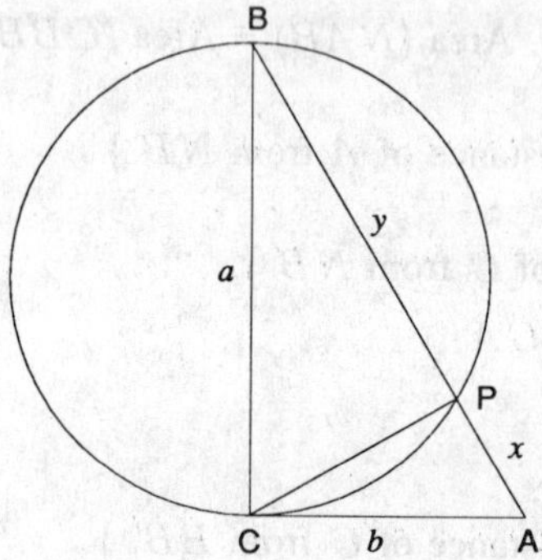


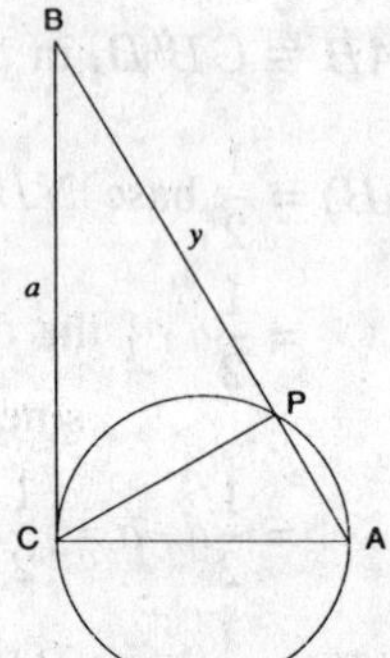


Again since $ABC \parallel\!| \, CBP$,

$$\frac{AB}{BC} = \frac{CB}{BP}, \frac{c}{a} = \frac{a}{y}, \underline{cy = a^2}$$

**Proof of the Hypotenuse Theorem [BHASKARA]**

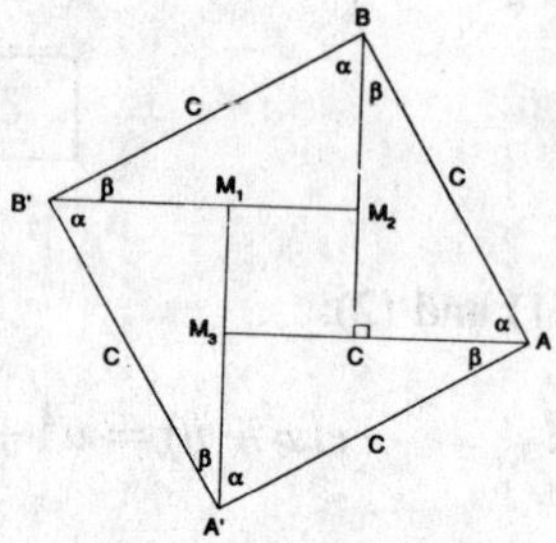


Draw $A'M_1 \parallel BC$ and $B'M_2 \perp BC$. We have that the four triangles $A'AM_3, BB'M_2, B'A'M_1, ABC$ are all congruent. If $\delta =$ Area $(ABC)$ then the sum of the areas of these four triangles is equal to

$$4 \cdot \delta = 4 \cdot \left(\frac{1}{2}ab\right) = 2ab$$

From the congruence $AA'M_3 \equiv ABC$ we get $A'M_3 = AC = b$, and from the congruence $B'A'M_1 \equiv ABC$ we get $A'M_1 = BC = a$, which yields:

$$M_1M_3 = A'M_1 - A'M_3 = a - b$$

Since

$$(ABB'A') = 4\delta + (CM_2M_1M_3)$$

we obtain

$$c^2 = 4\delta + (a-b)^2 = 2ab + (a-b)^2 = a^2 + b^2$$

**Proof of the Hypotenuse Theorem [Leonardo da Vinci]**

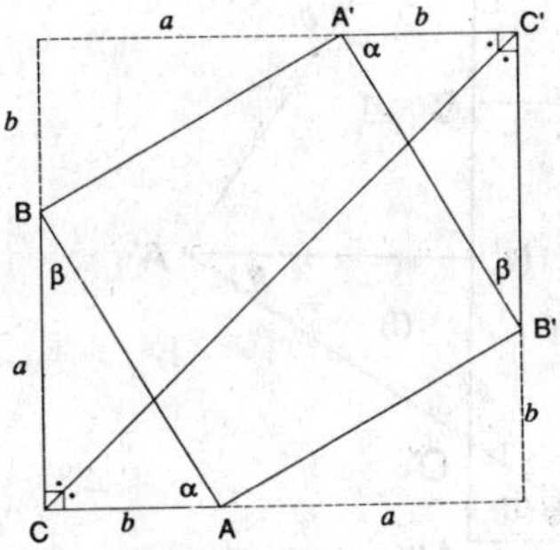


Fig.$1^0$

$$A'C' \parallel CA, B'C' \parallel CB$$

$$ABC \equiv A'B'C'$$

The distance of $C'$ from $BC$ is $= b + a =$ the distance of $C'$ from $CA$. Consequently $CC'$ bisects the angle $\widehat{ACB} = \frac{\pi}{2}$ We have the congruences:

(*) $A'BCC' \equiv ABM_1M_4$

$C'B'AC \equiv M_1M_2M_3M_4$

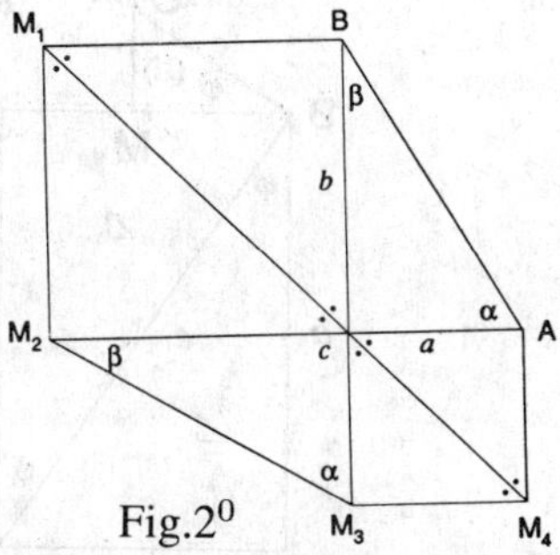


Fig.$2^0$

$M_1, C, M_4$ are collinear. The angles marked • are $= \frac{\pi}{4}$ From the two congruences (*) we get, by comparing their areas

$$(ABM_1M_4) + (M_1M_2M_3M_4)$$
$$= (A'BCC') + (C'B'AC)$$

i.e.,

$$(ABM_1M_2M_3M_4) = (A'BCAB'C')$$

$$a^2 + b^2 + 2(ABC) = c^2 + 2(ABC)$$

$$a^2 + b^2 = c^2$$

**Proof of the Hypotenuse Theorem [as in YUKTIBHĀSHĀ, the commentary on the TANTRA-SANGRAHA (1608)]**

We have: $BM = a - b = B'M'$

$$\frac{OM}{BM} = \tan\varphi = \frac{O'M'}{B'M'}$$

Hence $OM = OM'$ i.e.,

$\underline{OBM \equiv O'B'M'}$

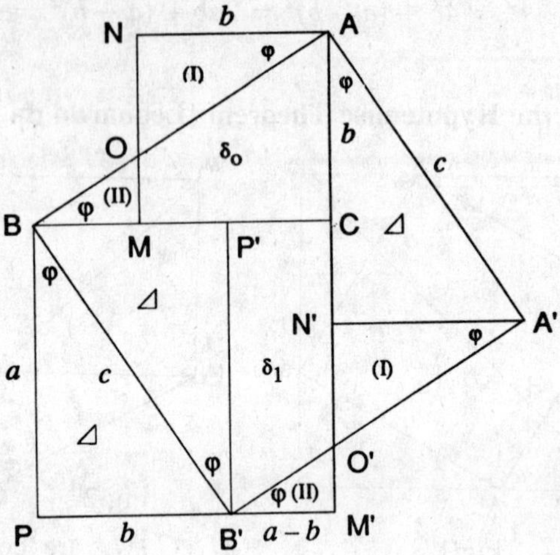


$$\Delta = (ABC)$$
$$\delta_0 = (OMCA)$$
Again, $$\delta_1 = (P'B'O'C)$$

(i) $\dfrac{ON}{AN} = \tan\varphi = \dfrac{O'N'}{A'N'}$

Since $AN = b$, and $A'N' = c\sin\varphi$ from the triangle $AA'N'$, while $\sin\varphi = \dfrac{AC}{BA} = \dfrac{b}{c}$ we get $A'N' = b$, thus

(ii) $AN = b = A'N'$

From (i) and (ii) it follows that $ON = O'N'$, thus $\underline{OAN \equiv O'A'N'}$ Also we have the congruences:

$$ABC \equiv B'BP \equiv BB'P' \equiv A'AN'$$

Accordingly,

$a^2$ = Square on $BC = 2\Delta + \delta_1$ + (II)
$b^2$ = Square on $AC = \delta_0$ + (I)
$c^2$ = Square on $AB = 2\Delta + \delta_0 + \delta_1$ + (I) + (II)

Consequently $c^2 = a^2 + b^2$. ■

We shall give yet another proof of the Hypotenuse Theorem.

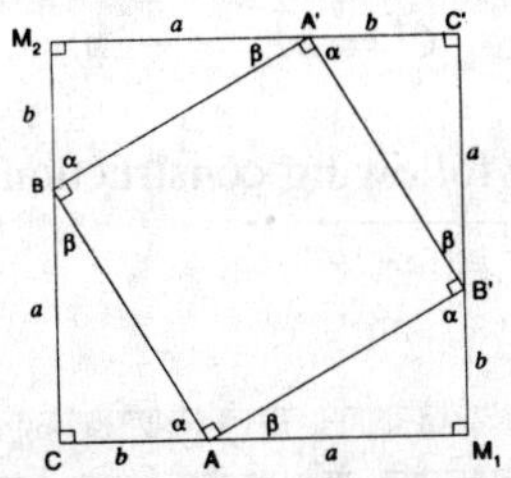

Draw the parallel through $A'$ to $CA$, and the parallel through $B'$ to $CB$. Then it is easy to see that

$$ABC \equiv B'AM_1 \equiv A'B'C' \equiv BA'M_2$$

Consequently
$a = BC = AM_1 = B'C' = A'M_2$
$b = CA = M_1B' = C'A' = M_2B$

Since

$$(CM_1C'M_2) = (AB'A'B') + 4\delta, \quad \delta = (ABC) = \frac{1}{2}ab$$

we obtain that

$$(a+b)^2 = c^2 + 2ab$$
$$a^2 + b^2 = c^2 \quad \blacksquare$$

**Theorem 2. [The Converse of the Hypotenuse Theorem]**

*If in a triangle* $ABC$ the sides $BC = a$, $CA = b$ and $AB = c$ satisfy the relation

$$c^2 = a^2 + b^2 \tag{3}$$

then $\hat{C} = \dfrac{\pi}{2}$

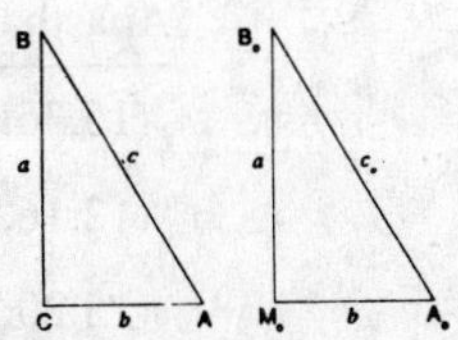

**Proof:** Construct a *right* triangle $A_0B_0M_0$ such that:
$M_0B_0 = a$, $M_0A_0 = b$
$\hat{M}_0 = \dfrac{\pi}{2}$
Then, if $c_0 = A_0B_0$, by Theorem 1,
$c_0^2 = a^2 + b^2$
By hypothesis (3), we therefore get $c_0^2 = c^2$, i.e. $c_0 = c$

Now that we have seen that $c_0 = c$, the two triangles $ABC$ and

$A_0B_0M_0$ are congruent. Consequently

$$\hat{C} = \hat{M}_0 = \frac{\pi}{2} \blacksquare$$

Kātyāyana gives the following construction for a right triangle:

यावत्रमाणानि समचतुरश्राण्येकीकर्तु चिकीर्षेत्
एकोनानि तानि भवन्ति तिर्यक् द्विगुणान्येकत
एकाधिकानि । त्र्यस्त्रि र्भवति तस्येषु स्तात्करोति ।

If we construct an isosceles triangle with base

$$BC = (n-1)a, \text{ and sides } AB = AC = \frac{n+1}{2}a$$

then the altitude $AP = \sqrt{n} \cdot a$ i.e., the square with side $AP$ will yield $n \cdot a^2$.

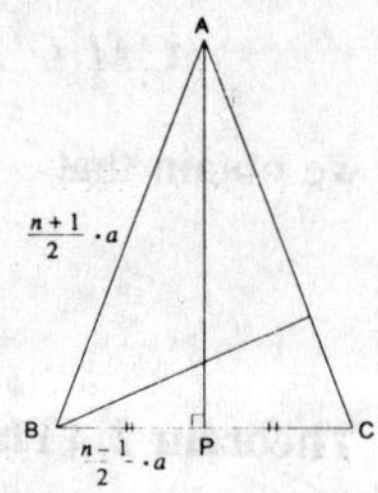


In fact, by the Hypotenuse theorem,

$$\begin{aligned} AP^2 &= AB^2 - BP^2 \\ &= \frac{(n+1)^2}{4}a^2 - \frac{(n-1)^2}{4}a^2 \\ &= n \cdot a^2 \end{aligned}$$

Such geometrical constructions were used in the construction of the *Védis*. The enlargement or the reduction of the same geometric pattern implied a knowledge of similar figures. In this manner Āpastamba and Bódhāyana list several right angles.

| Āpasthamba | Bandhāyana |
|---|---|
| (15,36,39) | (3,4,5) |
| (12,16,20) | (5,12,13) |
| (15,20,25) | (8,15,17) |
| (5,12,13) | (7,24,25) |
| (8,15,17) | (12,35,37) |
| (12,35,37) | (15,36,39) |

By taking $n = m^2$ in the triple

$$\left(\frac{n-1}{a}\cdot\sqrt{n}\cdot a,\ \frac{n+1}{2}a\right)$$

we can get the rational triangles in the form

$$\left(\frac{m^2-1}{2}a,\ ma,\ \frac{m^2+1}{2}a\right) \tag{4}$$

*Definition 1.* Three positive *integers* $(x, y, z)$, satisfying the equation

$$\boxed{x^2 + y^2 = z^2} \tag{5}$$

are said to form a *Pythagorean triple*

If $(x, y, z)$ is a Pythagorean triple, then, for every postive integer $k$, the triple $(kx, ky, kz)$ is again Pythagorean. A Pythagorean triple $(x, y, z)$ is called *primitive* if any two of $x, y$ and $z$ are coprime, i.e.

$$\begin{cases} (x,y) = (x,z) = (y,z) = 1 \\ x^2 + y^2 = z^2 \qquad\qquad (x > 0, y > 0, x > 0) \end{cases} \tag{6}$$

**Theorem 3.**

*The primitive solutions of the equations*

$$x^2 + y^2 = z^2,$$

*with $y$ even, are given by*

$$\boxed{(x, y, z) = (m^2 - n^2, 2mn, m^2 + n^2} \tag{7}$$

*where $m > 0, n > 0, (m, n) = 1$ and $m$ and $n$ are of opposite parity.*

**Proof**: Suppose $\{x, y, z\}$ is a solution in postive integers of the equation (5). Let $g = (x, y)$. Then $g^2|z^2$ which implies $g|z$. Since

$$(x, y, z) = ((x, y), z) = (g, z) = g,$$

[$(x, y)$, respectively $(x, y, z)$, denotes the g.c.d. of $x$ and $y$, respectively $x, y$ and $z$]. Let

$$\{x_0, y_0, z_0\} = \left\{\frac{x}{g}, \frac{y}{g}, \frac{z}{g}\right\} \text{ so that } \left|\ \begin{aligned} &(x_0, y_0) = (x_0, z_0) = (y_0, z_0) = 1, \\ &x_0^2 + y_0^2 = x_0^2, \end{aligned}\right.$$

i.e., $\{x_0, y_0, z_0\}$ is a primitive solution of (5). Conversely if $\{x_1, y_1, z_1\}$ is a primitive solution of (5), then, for every integer $k_1$ $\{kx_1, ky_1, kz_1\}$ is again a solution of (5). Accordingly, we need only determine the primitive solutions of (5).

Suppose, then, $\{x, y, z\}$ in a primitive solution of (5). thus

$$x^2 + y^2 = z^2 \text{ with } (x, y) = (x, z) = (y, z) = 1$$

Here $x$ and $y$ cannot *both* be even (for, otherwise, $2|x$ and $2|y$ contradicating $(x, y) = 1$). We assert that $x$ and $y$ cannot *both* be odd either. For, otherwise,

$$\left.\begin{matrix} x \equiv 1(\text{mod}2) \\ y \equiv 1(\text{mod}2) \end{matrix}\right\} \Rightarrow \left.\begin{matrix} x^2 \equiv 1(\text{mod}4) \\ y^2 \equiv 1(\text{mod}4) \end{matrix}\right\} \Rightarrow z^2 = x^2 + y^2 \equiv 2(\text{mod}4),$$

which is impossible because if $z$ is even $z^2 \equiv 0(\text{mod}4)$, while if $z$ is odd $z^2 \equiv 1(\text{mod}4)$. Hence $x$ and $y$ are of opposite parity. Since the equation (5) is symmetric in $x$ and $y$, we may assume, that:

(8) $\qquad x$ is odd, $y$ is even and $z$ is odd

We write the equation in the form

$$z^2 - x^2 = y^2 \text{ or } (z + x)(z - x) = y^2$$

since $z$ and $x$ are *both* odd, we must have herein $z + x$ and $z - x$ *both* even. Hence (8) implies

(8′) $\qquad z + x, z - x$ and $y$ are all even

Put

(9) $\qquad z + x = 2a, \quad z - x = 2b, \quad y = 2c;$

(since $x, y, z$ are positive integers and $z > x, a, b,$ and $c$ must be positive integers). Our equation $(z + x)(z - x) = y^2$ assumes the form $(2a) \cdot (2b) = (2c)^2$, i.e.,

(10) $\qquad ab = c^2$

From (9) we see that $a > b$, and since $(x, z) = 1$, it follows that $(a, b) = 1$. Now that $a$ and $b$ are coprime (10) is possible if and only if $a$ and $b$ are separately square integers:

$$a = m^2 \text{ and } b = n^2$$

whence, again from (10), $m^2n^2 = c^2$, or $c = mn$

Consequently, the primitive solutions of (5) are all given by

$$[x, y, z] = [m^2 - n^2, 2mn, m^2 + n^2]$$

where $m > n > o$ and $(m, n) = 1$, as claimed in (7). This completes the proof. ■

If $\{x_1, y_1, z_1\}$ and $\{x_2, y_2, z_2\}$ are Pythagorean triples, then $\{x_1x_2 - y_1y_2, x_1y_2 + x_2y_1, z_1z_2\}$ is again a Pythagorean triple.

In fact we have

$$(x_1x_2 - y_1y_2)^2 + (x_1y_2 + x_2y_1)^2 = (x_1^2 + y_1^2)(x_2^2 + y_2^2)$$
$$= z_1^2z_2^2$$

In particular, if $\{x, y, z\}$ is a Pythagorean triple, then $\{x^2 - y^2, 2xy, z^2\}$ is again a Pythagorean triple. Let us construct a sequence of Pytha go-rean triples as follows:

$$p_1 = \{x, y, z\}, p_2 = \{x^2 - y^2, 2xy, 2z\}, \cdots$$

and if $p_n = \{x_n, y_n, z_n\}$, then

$$p_{n+1} = \{xx_n - yy_n, xy_n + yx_n, zz_n\}$$

i.e.,

(11) $$x_{n+1} = xx_n - yy_n, y_{n+1} = xy_n + yx_n, z_{n+1} = zz_n$$

The points $P_n = \left(\frac{x_n}{z_n}, \frac{y_n}{z_n}\right)$ all lie on the unit circle $x^2 + y^2 = 1$. If $P = P_1 = \left(\frac{x}{z}, \frac{y}{z}\right) = (\cos 2\pi\alpha, \sin 2\pi\alpha)$, then

$$P_2 = (\cos 4\pi\alpha, \sin 4\pi\alpha), ..., \quad P_n = (\cos 2\pi n\alpha, \sin 2\pi n\alpha)$$

Accordingly, starting from the point $P_1$, we obtain a sequence of points all lying on the unit circle.

We recall:

$$\cos\frac{\pi}{4} = \frac{1}{\sqrt{2}}, \sin\frac{\pi}{4} = \frac{1}{\sqrt{2}}$$

(12) $$\cos\frac{\pi}{3} = \frac{1}{2}, \quad \sin\frac{\pi}{3} = \frac{\sqrt{3}}{2}$$

$$\cos\frac{\pi}{6} = \frac{\sqrt{3}}{2}, \sin\frac{\pi}{6} = \frac{1}{2}$$

$$\cos\frac{2\pi}{5} = \sin\frac{\pi}{10} = \frac{\sqrt{5}-1}{4}$$

$$\hat{B} = \hat{C} = \frac{2\pi}{5}, \hat{A} = \frac{\pi}{5}$$

$$\frac{AB}{BC} = \frac{BC}{CD}$$

$$\frac{1}{x} = \frac{x}{1-x}, x^2 + x - 1 = 0$$

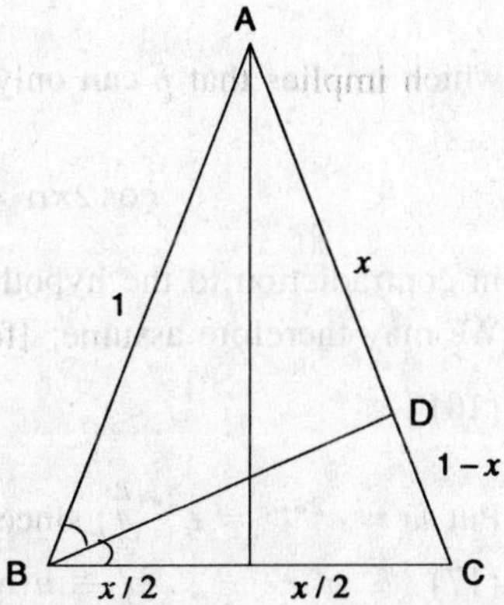

**Theorem 4. [E.HLAWKA (1979)]**

*If $\{x, y, z\}$ is any Pythagorean triple and*

(13) $$\boxed{\left(\frac{x}{y}, \frac{y}{z}\right) = (\cos 2\pi\alpha, \sin 2\pi\alpha)}$$

*then $\alpha$ is always* **irrational**.

Since $x, y, z$ are positive integers, our hypothesis means that

(14) $\cos 2\pi\alpha$ and $\sin 2\pi\alpha$ are *both rational*

We have to prove that

(15) (14) $\Rightarrow \alpha$ is *irrational*

Suppose, on the contrary, that $\alpha$ is rational:

(15′) $$\alpha = \frac{p}{q}, (p, q) = 1, q > 1$$

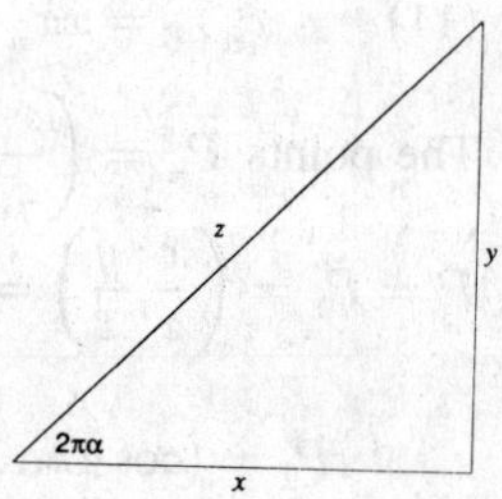


Since $2\pi\alpha < \dfrac{\pi}{2}$, we must have $\alpha < \dfrac{1}{4}$,

thus $\dfrac{p}{q} < \dfrac{1}{4}$; but then

$$\frac{1}{q} \leq \frac{p}{q} < \frac{1}{4}, \frac{1}{q} < \frac{1}{4}, q > 4,$$

i.e. $q$ is an integer $\geq 5$. Here, if $q = 5$, we have

$$\frac{p}{q} < \frac{1}{4}, p < \frac{q}{4} = \frac{5}{4},$$

which implies that $p$ can only be $= 1$. Thus $\alpha = \dfrac{p}{q} = \dfrac{1}{5}$. But then

$$\cos 2\pi\alpha = \cos \frac{2\pi}{5} = \frac{\sqrt{5} - 1}{4},$$

in contradiction to the hypothesis (14). Hence we cannot have $q = 5$. We may therefore assume: [for $q = 6$, See (12)]

(16) $$\boxed{q > 6}$$

Put $w = e^{2\pi i\alpha} = e^{2\pi i\frac{p}{q}}$; since $e^{2\pi ik} = 1$ for every integer $k$, we have

(17) $$n' \equiv n (\operatorname{mod} q) \Rightarrow w^{n'} = w^n,$$

in other words, the sequence $\{w^n\}$, $(n = 1, 2, \cdots)$, runs through the *finite* set

$$\{1, w, w^2, \ldots, w^{q-1}\}$$

i.e., for every $n = 1, 2, \cdots$ we can find $n'$, with $0 \leq n' \leq q - 1$ such that $w^n = w^{n'}$. Hence

$$z^q \cdot w^n = z^q \cdot w^{n'} = z^q \cdot \frac{(x+iy)^{n'}}{z^{n'}}, \quad \left(w = \frac{x+iy}{x}\right)$$
$$= z^{q-n'}(x+iy)^{n'}$$

Since $x, y, z$ are positive integers, we can write:

(18) $$z^q \cdot w^n = u_n + iv_n, \quad (u_n, v_n \text{ positive integers }) \quad (n = 1, 2, \ldots)$$

Therefore for every $k = 1, 2, \cdots$

$$(1 - w^n)^k = \sum_{r=0}^{k} (-1)^r \binom{k}{r} w^{nr}$$
$$= \sum_{r=0}^{k} (-1)^r \binom{k}{r} \frac{u_{mr} + iv_{nr}}{z^q}, \quad [See (18)]$$

Since $u_{nr}, v_{nr}$ are always *positive integers, [See (18)]*, it at once follows that

(19) $$\boxed{|1 - w^n|^k \geq \frac{1}{z^q}} \quad \text{for all } n, k$$

From the identity $|1 - e^{i\theta}|^2 = (1 - \cos\theta)^2 + \sin^2\theta = 2 - 2\cos\theta = 4\sin^2\frac{\theta}{2}$, we have:

$$|1 - w^n| = \left|1 - e^{2\pi i n \frac{p}{q}}\right| = 2\left|\sin\frac{\pi np}{q}\right|$$

Determine $p' > 0$ such that $pp' \equiv 1 (\bmod q)$; this is possible since, by (15'), $(p, q) = 1$. Then

$$\left|1 - w^{p'}\right| = 2\left|\sin\frac{\pi pp'}{q}\right| = 2\sin\frac{\pi}{q}$$

Owing to (16), we have $q > 6$, so that $\sin\frac{\pi}{q} < \sin\frac{\pi}{6} = \frac{1}{2}$. Thus

$$\left|1 - w_{p'}\right| < \frac{1}{2}$$

which implies that

$$\lim_{k\to\infty} \left|1 - w_{p'}\right|^k = 0$$

This is a contradiction to (19). Accordingly, our assumption (15') is wrong, and the assertion (15) is established. ■

## § 2. THE TRIANGLE

**Theorem 1. [HERON]**

If $2s$ is the perimeter of the triangle $ABC$ with sides $a, b, c$ then its area $\triangle = (ABC)$ is given by

$$\boxed{\triangle = \sqrt{s(s-a)(s-b)(s-c)}} \tag{1}$$

**Proof**: $x, y$ are called ≪*ābādhās*≫ of the sides $c, b$ respectively.

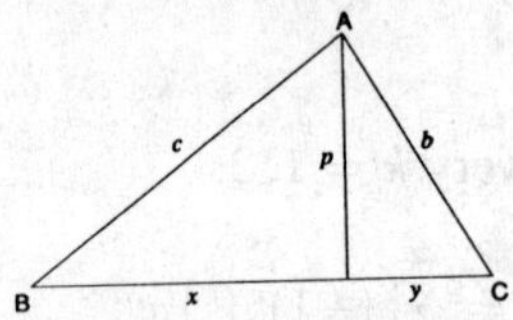


Since

$$p^2 = \left| \begin{array}{l} c^2 - x^2 \\ b^2 - y^2 \end{array} \right. \qquad BC = a = x + y$$

We have $c^2 - x^2 = b^2 - y^2$, so that $x^2 - y^2 = c^2 - b^2$. Since $x + y = a$, we get $a(x - y) = c^2 - b^2$, i.e.,

$$\begin{cases} x + y = a \\ x - y = \dfrac{c^2 - b^2}{a} \end{cases} \text{ whence follows } \begin{cases} x = \dfrac{1}{2}\left[a + \dfrac{c^2 - b^2}{a}\right] \\ y = \dfrac{1}{2}\left[a - \dfrac{c^2 - b^2}{a}\right] \end{cases}$$

Therefore

$$\begin{aligned} p^2 &= c^2 - x^2 = (c + x)(c - x) \\ &= \left(c + \frac{a^2 + c^2 - b^2}{2a}\right) \cdot \left(c - \frac{a^2 + c^2 - b^2}{2a}\right) \\ &= \frac{1}{4a^2}\{(a + c)^2 - b^2\} \cdot \{b^2 - (a - c)^2\} \\ &= \frac{1}{4a^2}(a + c + b)(a + c - b) \cdot (b + a - c)(b - a + c) \\ &= \frac{1}{4a^2} \cdot 2s \cdot (2s - 2b) \cdot (2s - 2c)(2s - 2a) \\ &= \frac{4}{a^2} \cdot s(s - a)(s - b)(s - c), \end{aligned}$$

Consequently $\Delta^2 = \dfrac{1}{4}a^2 \cdot p^2 = s(s - a)(s - b)(s - c)$. ■

**Example 1. [Bhaskara]**

*Given:* In the triangle $ABC$
$BC = \sqrt{18} - 1$
$CA = \sqrt{6}$
$AB = \sqrt{10} - \sqrt{5}$

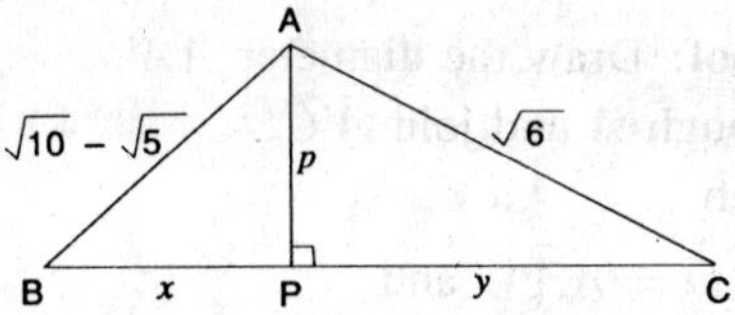


*To find:* $p$, the altitude of $A$ from $BC$.

We have

$$x + y = \sqrt{18} - 1 \text{ and } p^2 + x^2 = 10 + 5 - 2\sqrt{50} = 15 - 10\sqrt{2}$$
$$p^2 + y^2 = 6$$

so that

$$y^2 - x^2 = 6 - (15 - 10\sqrt{2}) = 10\sqrt{2} - 9$$
$$(y + x)(y - x) = 10\sqrt{2} - 9$$

$$y - x = \frac{10\sqrt{2} - 9}{y + x} = \frac{10\sqrt{2} - 9}{3\sqrt{2} - 1}$$
$$= \frac{(10\sqrt{2} - 9)(3\sqrt{2} + 1)}{(3\sqrt{2} - 1)(3\sqrt{2} + 1)}$$
$$= \frac{60 - 9 - 17\sqrt{2}}{18 - 1} = \frac{51 - 17\sqrt{2}}{17} = 3 - \sqrt{2}$$

Thus we have:

$$\left.\begin{aligned} x + y &= 3\sqrt{2} - 1 \\ -x + y &= 3 - \sqrt{2} \end{aligned}\right\} \text{ whence follows } \left.\begin{aligned} x &= 2\sqrt{2} - 2 \\ y &= \sqrt{2} + 1 \end{aligned}\right\}$$

Now

$$p^2 = 6 - y^2 = 6 - (\sqrt{2} + 1)^2 = 6 - (3 + 2\sqrt{2}) = 3 - 2\sqrt{2} = (\sqrt{2} - 1)^2$$

Consequently $\underline{p = \sqrt{2} - 1}$

**Theorem 2.**

त्रिभुजस्य वधो भुजयोद्विगुणितलम्बोध्दृतो हृदयरज्जुः ।
सा द्विगुणा त्रिचतुर्भुज कोण स्पृग्वृत्त विष्कम्भः ।।
ब्रह्मस्फुट सिध्दान्तः, XII, 27

*The product of the sides divided by twice the altitude is the radius of the circle that passes through the vertices of the triangle.*

**Proof:** Draw the diameter $AA'$ through $A$ and join $A'C$.
Then

$\widehat{ABD} = \widehat{AA'C}$ and

$\widehat{ADB} = \frac{\pi}{2} = \widehat{ACA'}$

so that $ABD \; ||| \; AA'C$. Hence

$\frac{AB}{AA'} = \frac{AD}{AC}$ i.e. $\frac{c}{2R} = \frac{p}{b}$

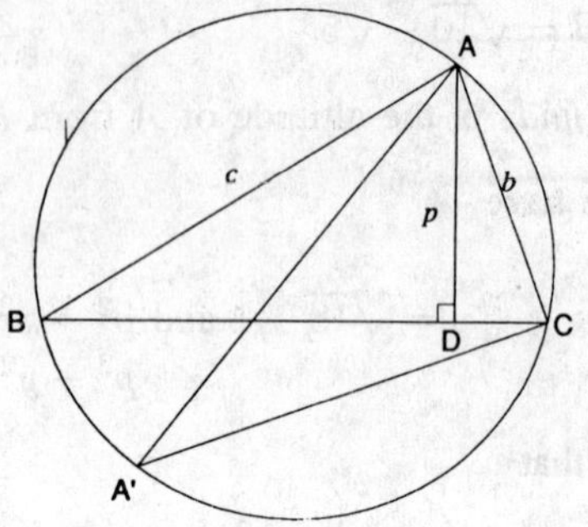


Since $\triangle = (ABC)$

$= \frac{1}{2} BC \cdot AD = \frac{1}{2} ap,$

we get

$\triangle = \frac{1}{2} a \cdot \frac{bc}{2R} = \frac{abc}{4R},$ $(AA' = 2R, R = \text{ circum radius })$

(2) $$\boxed{\triangle = \frac{abc}{4R}}$$

**Proof of (2) by Brahmagupta**

Denote by $M$ the middle point of the arc $\widehat{AC}$, and by $M'$ the diametrically opposite point. Then $M'$ will be the middle point of the complementary arc

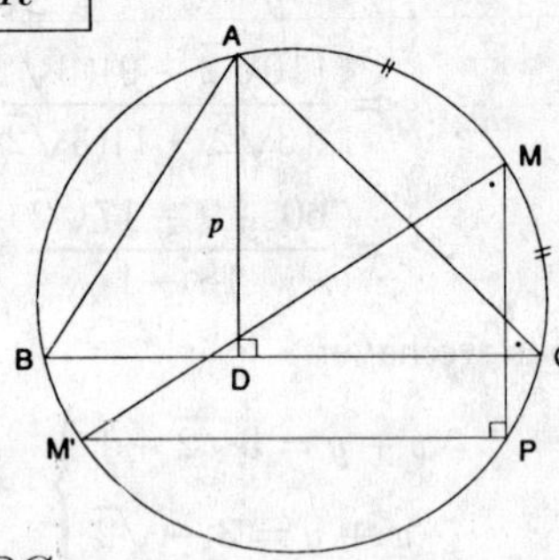


$$ABM'PC,$$

thus

$$\widehat{AB} + \widehat{BM}' = \widehat{M'P} + \widehat{PC}$$

Since $M'P \parallel BC$ we must have $\widehat{BM'} = \widehat{PC}$, so that

$$\widehat{AB} = \widehat{M'P} \text{ and thus } \widehat{ACB} = \widehat{M'MP}$$

Therefore $ACD \; ||| \; M'MP$, and accordingly

$$\frac{AC}{M'M} = \frac{AD}{M'P} \text{ i.e. } \frac{b}{2R} = \frac{p}{c}, \text{ since } M'P = AB = c$$

and thus

$$\triangle = \frac{1}{2}a \cdot p = \frac{1}{2}a \cdot \frac{bc}{2R} = \frac{abc}{4R} \blacksquare$$

The altitude $p = AD$ and the sides $AB = c$, $AC = b$ are connected by the relation

$$c \cdot \sin B = p = b \sin C$$

Similarly we get $a \sin C = c \sin A$, and $b \sin A = a \sin B$ Hence

(3) $$\frac{a}{\sin A} = \frac{b}{\sin B} = \frac{c}{\sin C}$$

From the right triangle $M'MP$ we see that

$$\frac{M'P}{M'M} = \sin \widehat{M'MP}, \text{i.e.,} \frac{c}{2R} = \sin C$$

Accordingly (3) can be written as

(4) $$\boxed{\frac{a}{\sin A} = \frac{b}{\sin B} = \frac{c}{\sin C} = 2R}$$

**Theorem 3.**

*If $P$ is any point, $a_1, a_2, a_3$ are the distances of $P$ from $A_1, A_2, A_3$ and $b_1, b_2, b_3$ are the distances of $P$ from the sides of the triangle $A_1A_2A_3$, then*

(5) $$\boxed{a_1 + a_2 + a_3 \geq 2(b_1 + b_2 + b_3)}$$

**Proof:** Denote by $\alpha_1, \alpha_2, \alpha_3$ the angles of the triangle $A_1A_2A_3$. From the triangle $A_1B_2B_3$ we see that

(6) $$\frac{B_2B_3}{\sin \alpha_1} = \text{diameter of the circle } A_1B_2B_3$$

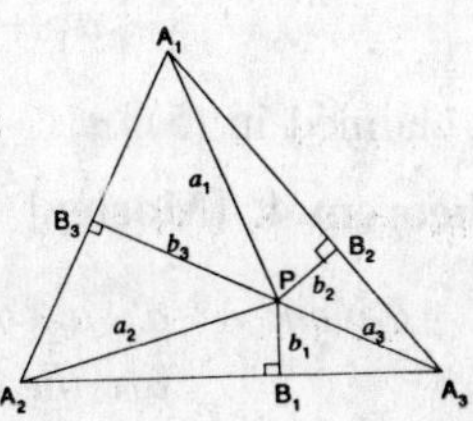


Since the quadrilateral $A_1B_2PB_3$ is cyclic, the diameter in (6) is $PA_1 = a_1$ hence (6) gives:

(7) $$B_2B_3 = a_1 \sin \alpha_1$$

Now,

the distance of $B_2$ from $B_1P = b_2 \cdot \sin\alpha_3$

the distance of $B_3$ from $B_1P = b_3 \cdot \sin\alpha_2$

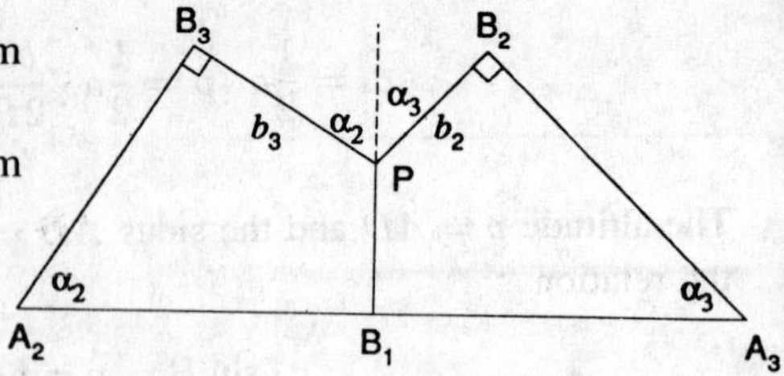


and we have,

[see figures (ii) and (iii)]

(8) $B_2B_3 \geq b_2 \sin\alpha_3 + b_3 \sin\alpha_2$

From (7) and (8) we have:

$a_1 \sin\alpha_1 \geq b_2 \sin\alpha_3 + b_3 \sin\alpha_2$

or

(9) $a_1 \geq b_2 \cdot \dfrac{\sin\alpha_3}{\sin\alpha_1} + b_3 \cdot \dfrac{\sin\alpha_2}{\sin\alpha_1}$

$$B_3\widehat{P}M_3 = \alpha_2$$
$$B_2\widehat{P}M_2 = \alpha_3$$
$$\begin{cases} B_2M_2 &= b_2 \sin\alpha_3 \\ B_3M_3 &= b_3 \sin\alpha_2 \end{cases}$$

Similarly we can obtain:

(9′) $$a_2 \geq b_3 \cdot \frac{\sin\alpha_1}{\sin\alpha_2} + b_1 \cdot \frac{\sin\alpha_3}{\sin\alpha_2}$$

(9″) $$a_3 \geq b_1 \cdot \frac{\sin\alpha_2}{\sin\alpha_3} + b_2 \cdot \frac{\sin\alpha_1}{\sin\alpha_3}$$

From (9), (9′), and (9″) it follows that

$$a_1 + a_2 + a_3 \geq b_1\left(\frac{\sin\alpha_2}{\sin\alpha_3} + \frac{\sin\alpha_3}{\sin\alpha_2}\right) + b_2\left(\frac{\sin\alpha_3}{\sin\alpha_1} + \frac{\sin\alpha_1}{\sin\alpha_3}\right) + b_3\left(\frac{\sin\alpha_1}{\sin\alpha_2} + \frac{\sin\alpha_2}{\sin\alpha_1}\right)$$

Since

$$\left(x + \frac{1}{x}\right) - 2 = \frac{1}{x}(x-1)^2 \geq 0 \text{ always, for } x > 0$$

i.e.,

$$x + \frac{1}{x} \geq 2, (x > 0),$$

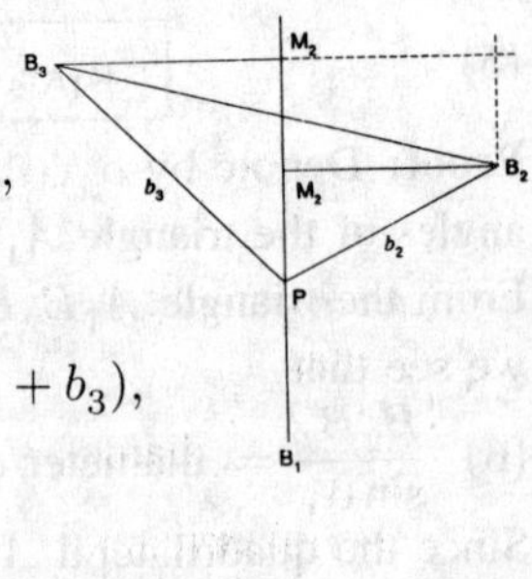


we conclude that

$$a_1 + a_2 + a_3 \geq 2(b_1 + b_2 + b_3),$$

as claimed in (5). ■.

**Theorem 4. [Morley]**

*Given*: $1^0$ $a_1, a_2, a_3$ *are the sides, and* $3\alpha_1, 3\alpha_2, 3\alpha_3$ *are the angles of a triangle* $A_1A_2A_3$

$2^0$ $M_1, M_2, M_3$ lie in the interior of the triangle

such that

$$M_1\widehat{A_2}A_3 = \alpha_2, \quad M_1\widehat{A_3}A_2 = \alpha_3$$
$$M_2\widehat{A_3}A_1 = \alpha_3, \quad M_2\widehat{A_1}A_3 = \alpha_1$$
$$M_3\widehat{A_1}A_2 = \alpha_1, \quad M_3\widehat{A_2}A_1 = \alpha_2$$

*Claim*:

$M_1M_2M_3$ is always equilateral.

**Proof**:

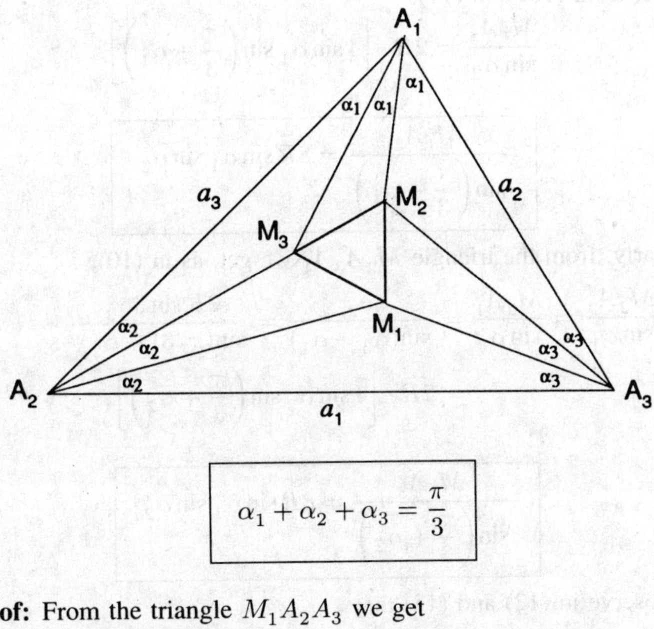


$$\boxed{\alpha_1 + \alpha_2 + \alpha_3 = \frac{\pi}{3}}$$

**Proof:** From the triangle $M_1A_2A_3$ we get

(10) $$\frac{M_1A_2}{\sin\alpha_3} = \frac{M_1A_3}{\sin\alpha_2} = \frac{a_1}{\sin(\alpha_2+\alpha_3)} = \frac{2R\sin 3\alpha_1}{\sin\left(\dfrac{\pi}{3}-\alpha_1\right)},$$

(where $R$ is the circumradius of the triangle $A_1A_2A_3$) Now we use a trigonometric identity proved in the following:

**Lemma 1**$^0$

(11) $$\boxed{\sin 3\theta = 4\sin\theta\sin\left(\frac{\pi}{3}+\theta\right)\sin\left(\frac{\pi}{3}-\theta\right)}$$

In fact,

$$4\sin\theta\sin\left(\frac{\pi}{3}+\theta\right)\sin\left(\frac{\pi}{3}-\theta\right) = 2\sin\theta\left(\cos 2\theta - \cos\frac{2\pi}{3}\right)$$
$$= 2\sin\theta\left(\cos 2\theta + \frac{1}{2}\right)$$
$$= 2\sin\theta\cos 2\theta + \sin\theta$$
$$= (\sin 3\theta - \sin\theta) + \sin\theta$$
$$= \sin 3\theta$$

Hence, from (10) and (11).

$$\frac{M_1A_3}{\sin\alpha_2} = 2R\cdot\left[4\sin\alpha_1\sin\left(\frac{\pi}{3}+\alpha_1\right)\right]$$

or

(12) $$\boxed{\frac{M_1A_3}{\sin\left(\frac{\pi}{3}+\alpha_1\right)} = 8R\sin\alpha_1\sin\alpha_2}$$

Similarly, from the triangle $M_2A_3A_1$ we get, as in (10),

$$\frac{M_2A_3}{\sin\alpha_1} = \frac{M_2A_1}{\sin\alpha_3} = \frac{a_2}{\sin(\alpha_3+\alpha_1)} = \frac{2R\sin 3\alpha_2}{\sin(\pi/3)-\alpha_2)}$$
$$= 2R\cdot\left[4\sin\alpha_2\sin\left(\frac{\pi}{3}+\alpha_2\right)\right],$$

thus

(13) $$\boxed{\frac{M_2A_3}{\sin\left(\frac{\pi}{3}+\alpha_2\right)} = 8R\sin\alpha_1\sin\alpha_2}$$

We observe in (12) and (13) that

$$\left(\frac{\pi}{3}+\alpha_1\right)+\left(\frac{\pi}{3}+\alpha_2\right)+\alpha_3 = \pi$$

We would like to conclude from (12) and (13) that

$$\frac{M_1M_2}{\sin\alpha_3} = 8R\sin\alpha_1\sin\alpha_2$$

For this purpose we need:

**Lemma $2^0$**

*Given* : In a triangle $ABC$ with sides $a, b, c$

(14) $$\frac{b}{\sin\beta} = \frac{c}{\sin\gamma} = \lambda$$

and $\hat{A} + \beta + \gamma = \pi$

*Claim*: $\dfrac{a}{\sin A} = \lambda, \hat{B} = \beta, \hat{C} = \gamma$

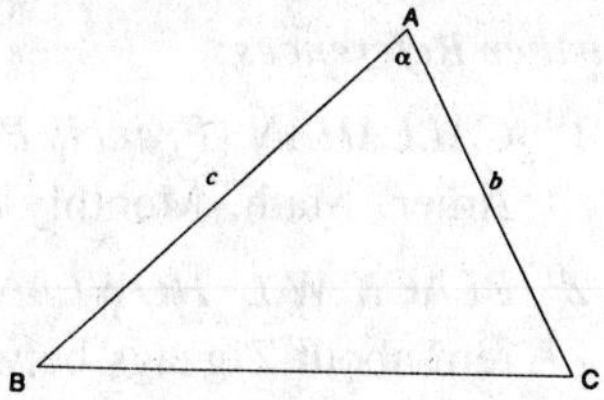


Put $\hat{A} = \alpha$; consider *any* triangle $A'B'C'$ with $\hat{A}' = \alpha, \hat{B}' = \beta, \hat{C}' = \gamma$

Then, from the triangle $A'B'C'$ we get

$$\frac{a'}{\sin\alpha} = \frac{b'}{\sin\beta} = \frac{c'}{\sin\gamma}$$

$$=: \lambda' \quad say$$

From (14),

$$b = \lambda \sin\beta = \lambda \cdot \left(\frac{b'}{\lambda'}\right)$$

$$c = \lambda \sin\gamma = \lambda \cdot \left(\frac{c'}{\lambda'}\right)$$

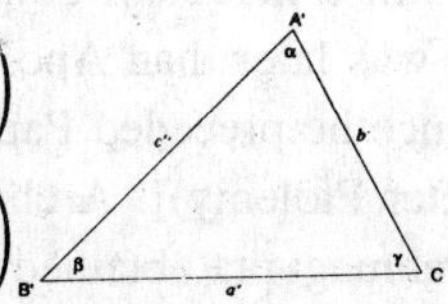


so that

$$\frac{b}{b'} = \frac{c}{c'} = \frac{\lambda}{\lambda'}$$

Now in the triangles $ABC$ and $A'B'C'$ we have

$$\hat{A} = \alpha = \hat{A}' \text{ and } \frac{b}{b'} = \frac{c}{c'} = \frac{\lambda}{\lambda'}$$

the triangles must therefore be similar. Hence

$$\hat{B} = \hat{B}' = \beta, \quad \hat{C} = \hat{C}' = \gamma$$

And

$$\frac{a}{a'} = \frac{b}{b'} = \frac{c}{c'} = \frac{\lambda}{\lambda'}$$

i.e.,

$$a = \frac{\lambda}{\lambda'} \cdot a' = \lambda \sin\alpha = \lambda \sin\hat{A}$$

Applying Lemma $2^0$ to our situation in (12) and (13), we have:

$$\frac{M_1 M_2}{\sin\alpha_3} = 8R \sin\alpha_1 \sin\alpha_2$$

or

$$M_1 M_2 = 8R \sin\alpha_1 \sin\alpha_2 \sin\alpha_3$$

By symmetry, we get the same expression for $M_2M_3, M_3M_1$:

$$\boxed{M_1M_2 = M_2M_3 = M_3M_1 = 8R \sin\alpha_1 \sin\alpha_2 \sin\alpha_3}$$

The Mortey's theorem is hereby established. ■.

*Further References*:

$1^0$ *CALLAHAN (Francis P.)*: ≪Morley Polygons≫
Amer. Math. Monthly *84* (1977), pp 325-337

$2^0$ *BLACK W.L. HOWLAND H.C., HOWLAND B)*: ≪A theorem about Zigzags between two circles≫
Amer. Math. Monthly 81 (1974) pp 754-757

## § 3. THE CYCLIC QUADRILATERAL

The formula for the are of a triangle in terms of its sides [*See (1) Th.1,* § *2. P.165*] was found by Heron of Alexandria [The date of Heron is still a matter of controversy. The possible limits are 150 B.C., (for he was later than Apollonius of Perga, and Hipparchus) and 250 A.D., (since he preceded Pappus). He is now placed in the 3rd century A.D. (after Ptolemy)]. Archimedes, it is sometimes said, knew this formula. Brahmagupta obtained a formula for the area of a *cyclic* quadrilateral in terms of its sides, and also expressions for its diagonals. This includes Heron's theorem as a special case; also the mode of derivation of the formula by Brahmagupta is entirely different from that of Heron. Brahmagupta's theorem about the diagonals of a cyclic quadrilateral was rediscovered by W.Snell in 1619 A.D.

Nārāyana Paṇḍita, the author of Gaṇita Kaumudi (1356 A.D.), obtains further results in the geometry of the cyclic quadrilateral:

सर्वचतुर्बाहूनां मुखस्य परिवर्तने यदा विहिते ।
$1^0$ कर्पस्तदा तृतीयः पर इति कर्णत्र्यं भवति ।।

When the top side and the flank side of a cyclic quadrilateral are interchanged, we get a third diagonal called ≪para≫ (the auxiliary diagonal).

$2^0$ द्विगुण व्यास विभक्ते त्रिकर्णघातेऽथवा गणितम् ।

The area of a cyclic quadrilateral is equal to the product of the three diagonals divided by twice the circum-diameter.

**Theorem 1. [Ptolemy]**

*Given*: $A_1, A_2, A_3, A_4$ *lie on a circle of radius* $R$

(1) $$A_2A_3 = a_1, A_3A_4 = a_2, A_4A_1 = a_3, A_1A_2 = a_4$$

The diagonals are

(2) $$d_{13} = A_1A_3 \text{ and } d_{24} = A_2A_4$$

*Claim:*

(3) $\boxed{a_1a_3 + a_2a_4 = d_{13}d_{24}}$

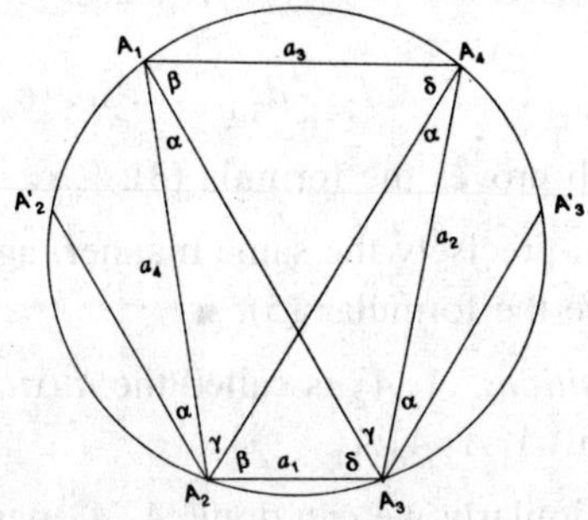


Draw $A_2A_2'$ parallel to the diagonal $A_2A_2'$., let

(4) $\boxed{d' = A_2'A_4}$

Then

(5) $\boxed{\begin{aligned} a_1a_2 + a_3a_4 &= d_{13}d' \\ a_1a_4 + a_2a_3 &= d_{24}d' \end{aligned}}$

The statement (3) is known as *Ptolemy's theorem*.

$$A_2A_2' \parallel A_3A_1 \text{ and } A_3A_3' \parallel A_2A_4$$

**Proof:** We read off from the figure:

$$A_1 = \alpha + \beta, A_2 = \beta + \gamma, A_3 = \gamma + \delta, A_4 = \delta + \alpha$$

$$A_1 + A_3 = \pi = A_2 + A_4$$

$$\varphi = \alpha + \gamma$$

We have

(6) $$2R = \frac{a_1}{\sin\alpha} = \frac{a_2}{\sin\beta} = \frac{a_3}{\sin\gamma} = \frac{a_4}{\sin\delta} =$$

$$= \frac{d_{13}}{\sin A_2} = \frac{d_{24}}{\sin A_1} = \frac{d'}{\sin\varphi}$$

Hence follows that

$$a_1a_3 + a_2a_4 = 4R^2(\sin\alpha\sin\gamma + \sin\beta\sin\delta) =$$
$$= 2r^2[\{\cos(\alpha-\gamma) - \cos(\alpha+\gamma)\} + \{\cos(\beta-\delta) - \cos(\beta+\delta)\}]$$

Since here $\cos(\alpha+\gamma) = \cos\varphi$ and $\cos(\beta+\delta) = \cos(\pi-\varphi) = -\cos\varphi$ we obtain that

$$a_1a_2 + a_2a_4 = 2r^2[\cos(\alpha-\gamma) + \cos(\beta-\delta)]$$

$$= 4r^2\cos\frac{\alpha-\gamma+\beta-\delta}{2}\cos\frac{\alpha-\gamma-\beta+\delta}{2}$$

$$= 4r^2\cos\frac{A_1-A_3}{2}\cos\frac{A_4-A_2}{2}$$

$$= 4r^2\cos\left(A_1 - \frac{\pi}{2}\right)\cos\left(\frac{\pi}{2} - A_2\right),$$

$$\text{since } A_1 + A_3 = \pi = A_2 + A_4$$

$$= 4r^2 \sin A_1 \sin A_2$$

$$= d_{24} \cdot d_{13}, \qquad see(6),$$

which proves the formula (3).

In precisely the same manner, again using the relations (6), we can derive the formulas (5). ■

*Definition:* $A_2A_2'$ is called the *auxiliary diagonal* of the cyclic quadrilateral $A_1A_2A_3A_4$

Similarly we can draw $A_3A_3'$ parallel to the diagonal $A_2A_4$. Then,

$$(6) \qquad \boxed{A_3'A_1 = 2R\sin\varphi = A_2'A_4 = d'}$$

*Remark:*

We can apply the formula (3) to the quadrilateral $A_1A_2'A_3A_4$; thus

$$(7) \qquad A_1A_2' \cdot A_3A_4 + A_1A_4 \cdot A_2'A_3 = A_1A_3 \cdot A_2'A_4$$

Herein, we observe that, (recall $A_2'A_2 \parallel A_1A_3$)

$$A_1A_2' = A_2A_3 = a_1, A_2'A_3 = A_1A_2 = a_4,$$

and $A_2'A_4 = 2R\sin(\alpha+\gamma) = 2R\sin\varphi = d'$. Hence (7) becomes

$$a_1a_2 + a_3a_4 = d_{13}d'$$

which yields the first relation in (5)

Similarly, we can apply the formula (3) to the quadrilateral $A_1A_2A_3'A_4$; thus

$$(8) \qquad A_1A_2 \cdot A_3'A_4 + A_1A_4 \cdot A_2A_3' = A_1A_3' \cdot A_2A_4$$

Since,recall $(A_3A_3' \parallel A_2A_4)$,

$$A_3'A_4 = A_2A_3 = a_1, \quad A_2A_3' = A_3A_4 = a_2,$$
$$A_1A_3' = 2R\sin(\alpha+\gamma) = 2R\sin\varphi = d',$$

it follows from (8) that

$$a_4a_1 + a_3a_2 = d' \cdot d_{24},$$

which yields also the second relation in (5). ■

Eliminating $d'$ in (5) we get

$$(9) \qquad \boxed{\frac{a_1a_2 + a_3a_4}{a_1a_4 + a_2a_3} = \frac{d_{13}}{d_{24}}}$$

From (3) and (9) we obtain:

**Theorem 2. [Brahmagupta]**

The diagonals of a cyclic quadrilateral can be expressed in terms of its sides:

$$\boxed{d_{13}^2 = \frac{(a_1a_2 + a_3a_4)(a_1a_3 + a_2a_4)}{(a_1a_4 + a_2a_3)}}$$

*and*

(10) $$\boxed{d_{24}^2 = \frac{(a_1a_3 + a_2a_4)(a_1a_4 + a_2a_3)}{(a_1a_2 + a_3a_4)}}$$

If $(A_1A_2A_3A_4)$ denotes the are of the cyclic quadrilateral we have

$$(A_1A_2A_3A_4) = (A_1A_2A_4) + (A_3A_2A_4)$$

By the relation (2), [See p. 170],

$$(A_1A_2A_4) = \frac{a_3a_4d_{24}}{4R}$$

$$(A_3A_2A_\triangle) = \frac{a_1a_2d_{24}}{4R}$$

Consequently

$$\triangle = (A_1A_2A_3A_4) = \frac{d_{24}}{4R}(a_1a_2 + a_3a_4)$$

$$= \frac{d_{24}}{4R} \cdot d_{13}d', \quad [See (5)]$$

Hence we obtain, [See also the statement $2^0$ in p. 176].

**Theorem 3. [Narayana Pandita (1356 A.D.)]**

The area of a cyclic quadrilateral is given by

(11) $$\boxed{\triangle = \frac{d_{13}d_{24}d'}{4R}}$$

We shall now prove:

**Theorem 4. [Brahmagupta]**

*The area of a cyclic quadrilateral is given in terms of its sides by*

(12) $$\boxed{\triangle = \sqrt{(\sigma - a_1)(\sigma - a_2)(\sigma - a_3)(\sigma - a_4)}}$$

*where* $2\sigma = a_1 + a_2 + a_3 + a_4$.

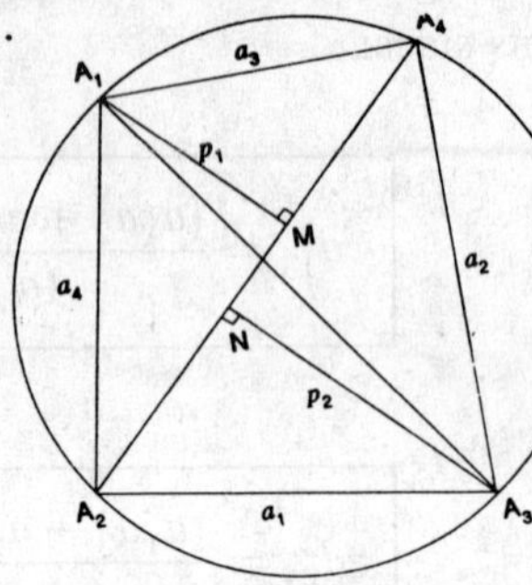


**Proof:** Drop perpendiculars $A_1M$ and $A_3N$ on the diagonal $A_2A_4$, and let $p_1 = A_1M$ and $p_2 = A_3N$

Then

$$(A_1A_2A_4) = \frac{1}{2}p_1 \cdot d_{24}$$

$$(A_3A_2A_4) = \frac{1}{2}p_2 \cdot d_{24}$$

so that

(13) $$\triangle = (A_1A_2A_3A_4) = \frac{1}{2}(p_1 + p_2) \cdot d_{24}$$

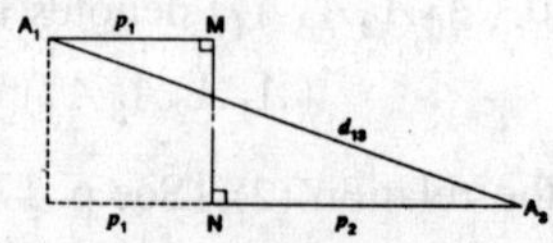


Since

$a_4^2 = p_1^2 + A_2M^2$ and $a_3^2 = p_1^2 + A_4M^2$

we have

(14) $$a_4^2 - a_3^2 = A_2M^2 - A_4M^2 = (A_2M + A_4M)(A_2M - A_4M) = d_{24} \cdot (A_2M - A_4M)$$

Similarly

$$a_1^2 = p_1^2 + A_2N^2 \text{ and } a_2^2 = p_2^2 + A_4N^2,$$

so that

(15) $$a_2^2 - a_1^2 = A_4N^2 - A_2N^2 = (A_4N + A_2N)(A_4N - A_2N) = d_{24} \cdot (A_4N - A_2N)$$

From (14) and (15) follows:

$$\begin{aligned}(a_4^2 - a_3^2) + (a_2^2 - a_1^2) &= d_{24}\{(A_2M - A_4M) + (A_4N - A_2N)\}\\ &= d_{24}\{(A_2M - A_4N) + (A_4N - A_4M)\}\\ &= d_{24} \cdot 2MN\end{aligned}$$

thus

(16) $$MN = \frac{-a_1^2 + a_2^2 - a_3^2 + a_4^2}{2d_{24}}$$

It can be easily seen that

(17) $$d_{13}^2 = MN^2 + (p_1 + p_2)^2$$

From (13), (16) and (17) we have:

$$\begin{aligned}
\triangle^2 &= \frac{1}{4}\cdot d_{24}^2(d_{13}^2 - MN^2)\\
&= \frac{1}{4}\left[d_{13}^2 d_{24}^2 - \frac{(-a_1^2+a_2^2-a_3^2+a_4^2)^2}{4}\right]\\
&= \frac{1}{16}[4(a_1a_3+a_2a_4)^2 - (a_1^2-a_2^2+a_3^2-a_4^2)^2],
\end{aligned}$$

See (3). p. 177 Ptolemy's theorem]

$$\begin{aligned}
&= \frac{1}{16}[\{2(a_1a_3+a_2a_4)+(a_1^2-a_2^2+a_3^2-a_4^2)\}\cdot\\
&\qquad\cdot\{2(a_1a_3+a_2a_4)-(a_1^2-a_2^2+a_3^2-a_4^2)\}]\\
&= \frac{1}{16}\{(a_1+a_3)^2-(a_2-a_4)^2\}\cdot\{(a_2+a_4)^2-(a_1-a_3)^2\}\\
&= \frac{1}{16}(a_1+a_3+a_2-a_4)(a_1+a_3-a_2+a_4)\cdot\\
&\qquad(a_2+a_4+a_1-a_3)(a_2+a_4-a_1+a_3)\\
&= \frac{1}{16}(2\sigma-2a_4)(2\sigma-2a_2)\cdot(2\sigma-2a_3)(2\sigma-2a_1)\\
&= (\sigma-a_1)(\sigma-a_2)(\sigma-a_3)(\sigma-a_4),
\end{aligned}$$

which establishes (12) and completes the proof of the theorem. ∎

From Theorem 3 and Theorem 4, [See (11) and (12)], we can derive an expression for the circumradius of a cyclic quadrilateral in terms of its sides:

**Theorem 5. [Paramesvara (1430 A.D.)]**

The circumradius $R$ of a cyclic quadrilateral with sides $a_1, a_2, a_3, a_4$ and perimeter $2\sigma$ is given by

$$R = \frac{\sqrt{(a_2a_3+a_1a_4)(a_3a_1+a_2a_4)(a_1a_2+a_3a_4)}}{\sqrt{(\sigma-a_1)(\sigma-a_2)(\sigma-a_3)(\sigma-a_4)}} \tag{18}$$

**Proof:** We have, from (3) and (5),

$$(18') \qquad \boxed{\begin{aligned} a_1a_3 + a_2a_4 &= d_{13}d_{24} \\ a_1a_2 + a_3a_4 &= d_{13}d' \\ a_1a_4 + a_2a_3 &= d_{24}d' \end{aligned}}$$

By multiplying the three equations in (18') we obtain

$$\begin{aligned}(a_2a_3 + a_1a_4)(a_3a_1 + a_2a_4)(a_1a_2 + a_3a_4) &= \\ = (d_{13}d_{24}).(d_{13}d').(d_{24}d' &= (d_{13}^2 d_{24}^2.d'^2\end{aligned}$$

Since by (11),

$$R = \frac{d_{13}d_{24}d'}{4\triangle},$$

where $\triangle$ is given by (12), we obtain (18). ■

The formula of Paramesvara was rediscovered in Europe by Lhuiller in 1782.

## § 4. THE CIRCLE

In this section we use the geometry of the complex plane. We recall that:

$$0, z_1, z_1 + z_2, z_2$$

will always form a parallelogram. Hence the ray $\overrightarrow{0, z_1 - z_0}$ is equal and parallel to the ray $\overrightarrow{z_0, z_1}$. Denote by $\varphi$ the angle that $\overrightarrow{z_0 z_1}$ makes with $\overrightarrow{ox}$. If $d = |z_1 - z_0|$ we can write

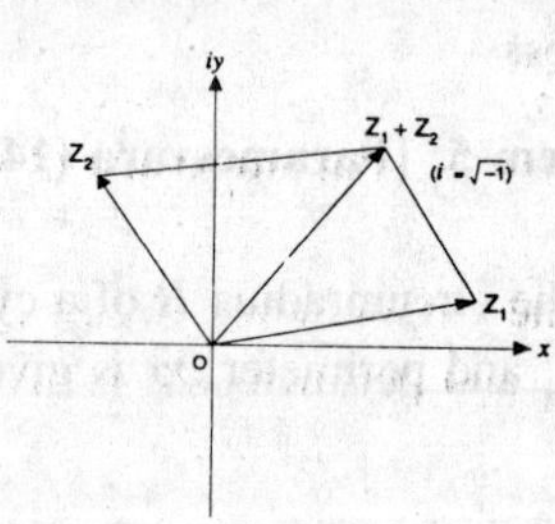


$$(1) \qquad \boxed{z_1 - z_0 = de^{i\varphi},} \qquad (0 \leq \varphi < 2\pi)$$

Herein
$\varphi = \text{Arg}(z_1 - z_0)$

If

(2) $\begin{cases} z_1 - z_0 = a_1 e^{i\varphi_1} \\ z_2 - z_0 = a_2 e^{i\varphi_2} \end{cases}$

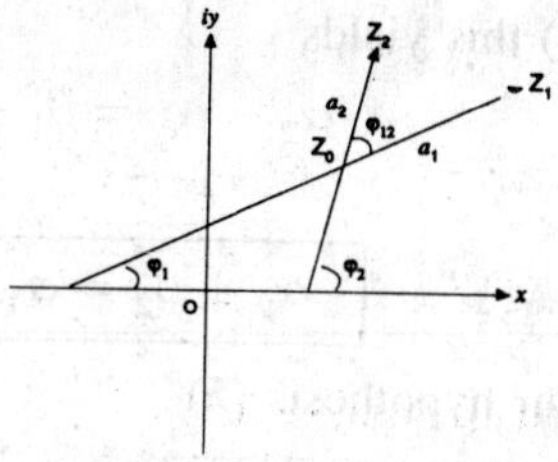


then

$$\frac{z_2 - z_0}{z_1 - z_0} = \frac{a_2}{a_1} \cdot e^{i(\varphi_2 - \varphi_1)}$$
$$= \frac{a_2}{a_1} \cdot e^{i\varphi_{12}}$$

i.e.,

$$\underline{\varphi_{12} = \varphi_2 - \varphi_1}$$

(3) $$\boxed{\varphi_{12} = \text{Arg}\,\frac{z_2 - z_0}{z_1 - z_0} = \sphericalangle\,(\overrightarrow{z_0 z_1}, \overrightarrow{z_0 z_2})}$$

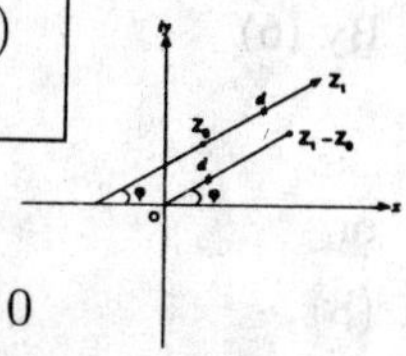


We have the obvious indentities:

$$(z_2 - z_3) + (z_3 - z_1) + (z_1 - z_2) = 0$$
$$z_1(z_2 - z_3) + z_2(z_3 - z_1) + z_3(z_1 - z_2) = 0$$

which imply the identity

(4)
$$\boxed{(z_0 - z_1)(z_2 - z_3) + (z_0 - z_2)(z_3 - z_1) + (z_0 - z_3)(z_1 - z_2) = 0}$$

Let us now make the following hypothesis:

(5) *Given:* $z_0, z_1 \cdot z_2, z_3$ *concyclic*

Put

(6)
$$\begin{array}{l|l} z_1 - z_0 = a_1 e^{i\alpha_1} & z_2 - z_3 = b_1 e^{i\delta_1} \\ z_2 - z_0 = a_2 e^{i\alpha_2} & z_3 - z_1 = b_2 e^{i\beta_2} \\ z_3 - z_0 = a_3 e^{i\alpha_3} & z_1 - z_2 = b_3 e^{i\beta_3} \end{array}$$

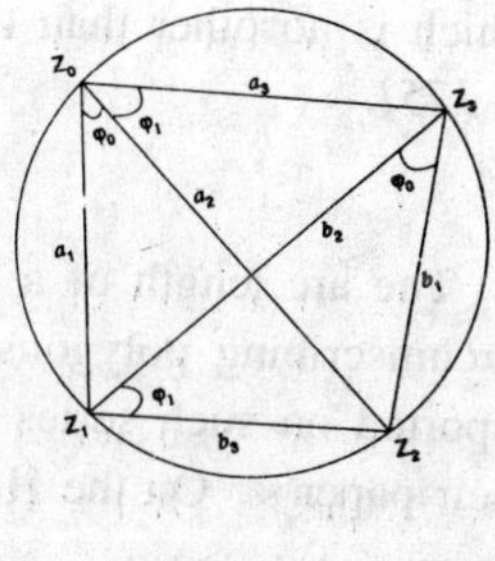


$(0 \leq \alpha_i, \beta_i < 2\pi, i = 1, 2, 3)$

Because of our hypothesis (5)

$(\overrightarrow{z_0 z_1}, \overrightarrow{z_1 z_2}) = \varphi_0 \cdot (\overrightarrow{z_3 z_1}, \overrightarrow{z_3 z_2})$

By (3) this becomes

$$\text{Arg}\,\frac{z_2 - z_0}{z_1 - z_0} = \text{Arg}\,\frac{z_2 - z_3}{z_1 - z_3}$$

i.e.,

$$\text{Arg}(z_2 - z_0) - \text{Arg}(z_1 - z_0) = \text{Arg}(z_2 - z_3) - \text{Arg}(z_1 - z_3)$$

Owing to (6) this yields

$$\alpha_2 - \alpha_1 = \beta_1 - (\beta_2 + \pi)$$

or

(7) $$\boxed{\alpha_2 + \beta_2 = \alpha_1 + \beta_1 - \pi}$$

Again, by our hypothesis (5)

$$(\overrightarrow{z_0 z_2}, \overrightarrow{z_0 z_3}) = \varphi_1 = (\overrightarrow{z_1 z_2}, \overrightarrow{z_1 z_3})$$

which, again by (3) gives

$$\text{Arg}(z_3 - z_0) - \text{Arg}(z_2 - z_0) = \text{Arg}(z_3 - z_1) - \text{Arg}(z_2 - z_1)$$

By (6)

$$\alpha_3 - \alpha_2 = \beta_2 - (\beta_3 + \pi)$$

or

(8) $$\boxed{\alpha_3 + \beta_3 = \alpha_2 + \beta_2 - \pi}$$

Substituting from (6) in the identity (4), we have:

$$a_1 b_1 e^{i(\alpha_1+\beta_1)} + a_2 b_2 e^{i(\alpha_2+\beta_2)} + a_3 b_3 e^{i(\alpha_3+\beta_3)} = 0$$

Consequently, it follows from (7) and (8) that

$$e^{i(\alpha_2+\beta_2)}[a_1 b_1 e^{i\pi} + a_2 b_2 + a_3 b_3 e^{-i\pi}] = 0$$

Since $e^{i\pi} = e^{-i\pi} = -1$ we obtain, finally,

$$\boxed{a_1 b_1 + a_3 b_3 = a_2 b_2}$$

which is no other than the statement of Ptolemy's theorem *[See Th.1 P. 175]*.

### Series for $\pi$

The arc length of a circle can be approximated by inscribing and circumscribing polygons. C.T. Rajagopal and Mukunda Marar have reported on such series (along with the original texts in Sanskrit) in their paper ≪ On the Hindu quadrature of the circle.≫

The sine and the cosine series

$$\sin x = x - \frac{x^3}{3!} + \frac{x^5}{5!} - + \cdots$$

$$\cos x = 1 - \frac{x^2}{2!} + \frac{x^4}{4!} - + \cdots$$

were well known in the Āryabhaṭa school. They are derived in Yukti-Bhāsha by Mādhava (14th Century). This derivation is based on a method known as ≪SAMKALITA≫ which anticipates ideas of Differential and Integral calculus. Madhava had also discovered the series

$$\tan^{-1} t = t - \frac{t^3}{3} + \frac{t^5}{5} + \cdots$$

which is known as ≪Gregory's Series≫ (1671) in England, and as ≪Leibnitz series≫ in Germany. For further details the reader may consult the references.

$1^0$ *T.A. Sarasvati Amma:* ≫Geometry in Ancient and Medieval India≫, Chapter VIII, § § 7.10 and 7.11, PP 173-190

$2^0$ *Charles M. Whish:* ≪On the Hindu quadrature of the Circle, and the infinite series in the four Śāstras, the Tantra Sangraham, Yukti Bhāshā, Carana Padhati, and Ṣaḍratnamālā (Royal Asiatic Society, Read the 15th December 1832)

Nîlakanṭha Sômayājin quotes a result of Mādhava:

विबुधनेत्रगजाहि हुताशन त्रिगुणवेदभवारण बाहवः ।
नवनिखर्वमितेवृतिविस्तरे परिधिमानमिदं जगदुर्बुधाः ।

(Kriyā-Kramakāri)

The ratio of the circumference of a circle of diameter = $9.10^{11}$ is 2 827 433 388 233.

i.e.,

$$\pi = \frac{2\ 827\ 433\ 388\ 233}{9.10^{11}} = 3.1415\ 9265\ 359\cdots$$

*(Mādhava's rational approximation to $\pi$)*

In *Karaṇa Paḍḍhati* and *Ṣadratnamāla* there are two verses which yield the following approximations to $\pi$.

**(1)** ≪If the circumference of a circle in minutes be multiplied by $10^{10}$, and the product be divided by 31 41 59 26 536, the quotient will be the diameter of the circle in terms of the minutes of the circumference, and its half will be the radius.≫

**(2)** ≪If you proceed thus (as laid down in the former verse), and measure the circumference of a great circle by $10^{17}$, the circumference will be equal to 3 1415 9265 3589 79324 $\cdots$

Charles M. Whish states:

≪The approximations to the true value of the circumference with a given diameter, exhibited in these three works, are so wonderfully correct, that European mathematicians, who seek for such proportion in the doctrine of fluxions, or in the more tedious continual bisection of an arc, will wonder by what means the Hindu has been able to extend the proportion to so great a length.≫

In *Tantra Sangraham* the infinite series are given for $\pi$: with a correction for the error term:

$$(9) \qquad C \simeq 4d - \frac{4d}{3} + \frac{4d}{5} - + \cdots \pm \frac{4d.\left(\frac{1}{2}p\right)}{p^2+1}$$

($p$ = last odd number - 1)

$$(10) \qquad C \simeq 4\left[1 - \frac{1}{3} + \frac{1}{5} - + \cdots \pm \frac{\frac{p^2}{4}+1}{(p^2+4+1)\frac{p}{2}}\right], \quad \underline{(d=1)}$$

(p = last odd number +1)

In *Karana-Padhati* other series have been described

$$(11) \qquad C = d\left[3 + \frac{4}{3^3-3} - \frac{4}{5^2-5} + \frac{4}{7^2-7} - + \cdots\right]$$

which yields

$$(12) \qquad \pi = 3 + 4\left[\frac{1}{2\cdot 3\cdot 4} - \frac{1}{4\cdot 5\cdot 6} + \frac{1}{6\cdot 7\cdot 8} - + \cdots\right]$$

$$\pi = 2 + 4\left[\frac{1}{1\cdot 3} - \frac{1}{3\cdot 5} + \frac{1}{5\cdot 7} - + \cdots \pm \frac{1}{[(p-1)^2+2]\cdot 2}\right]$$

The author realizes that the series converge very slowly, and provides for correction terms which yield remarkably close approximations to $\pi$.

### Mādhava's Series for $\pi$

Let us consider the unit circle $\Gamma$. Let $\theta$ be an angle such that $|\theta| \leq \frac{\pi}{4}$ so that $t = \tan\theta$ has $|t| \leq 1$.

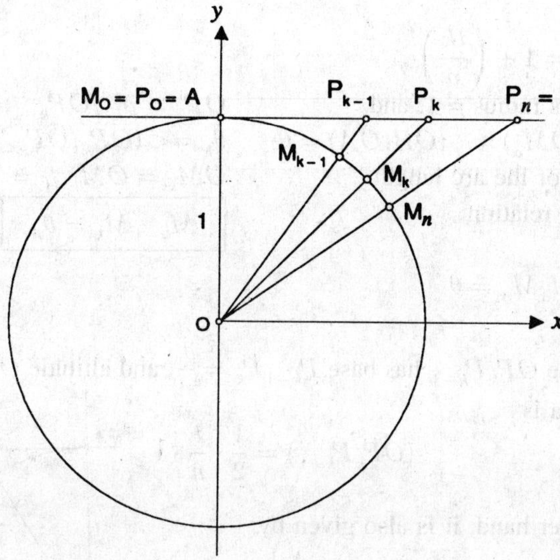


$\theta =< (\overrightarrow{OB}, \overrightarrow{Oy})$

$\theta_k =< (OP_k, OP_{k-1})$

$t = \tan\theta$

$|\theta| \leq \frac{\pi}{4} \cdot |t| \leq 1$

From the triangle $AOB$ we see that $t = \tan\theta = \tan \widehat{AOB} = \frac{AB}{AO} ==$ $AB$. Divide $AB$ into $n$ equal parts:

$$AP_1, P_1P_2, \cdots P_{k-1}P_k, \cdots, P_{n-1}B \quad \text{(with } P_0 = A \text{ and } P_n = B\text{)}$$

Then

$$(14) \qquad P_{k-1}P_k = \frac{AB}{n} = \frac{\tan\theta}{n} = \frac{t}{n} \quad (k = 1, 2, \cdots n)$$

Let $OP_k$ meet $\Gamma$ in $M_k$, and put $\rho_k = OP_k \ (k = 0, 1, 2, \ldots, n)$

Since $OA^2 + AP_k^2 = OP_k^2$,
we have

(16) $\rho_k^2 = 1 + \left(\frac{tk}{n}\right)^2$

Since $\Gamma$ has radius = 1, and $(OM_n, OM_o) = \;(OB, OA) = \theta$
We have, for the arc length $M_o\widehat{M}_n$ the relation

$OP_k = \rho_k,\ OP_{k-1} = \rho_{k-1}$
$\theta_k =< (OP_k, OP_{k-1})$
$OM_k = OM_{k-1} = 1$

$$\boxed{M_{k-1}\widehat{M}_k = \theta_k}$$

(17) $M_o\widehat{M}_n = \theta$

The triangle $OP_kP_{k-1}$ has base $P_{k-1}P_k = \frac{t}{n}$, and altitude $OA = 1$, so that its area is

$$(OP_kP_{k-1}) = \frac{1}{2} \cdot \frac{t}{n} \cdot 1$$

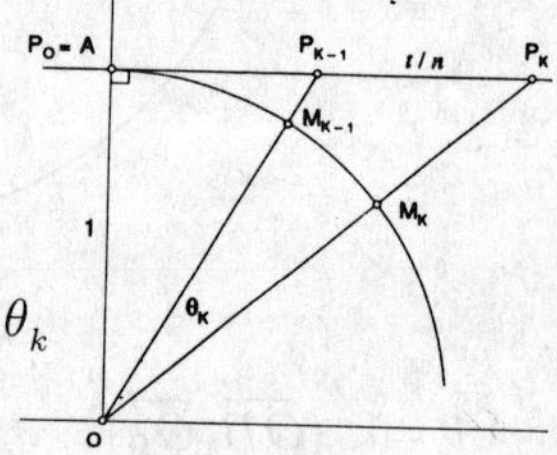


On the other hand, it is also given by

$$(OP_kP_{k-1}) = \frac{1}{2}\rho_k\rho_{k-1}\sin\theta_k$$

Hence follows

(18) $$\sin\theta_k = \frac{t}{n} \cdot \frac{1}{\rho_k\rho_{k-1}}, \quad (k = 1, \cdots n)$$

Since $\frac{\rho_k^2}{\rho_{k-1}^2} \to 1$ as $n \to \infty$, we have $\rho_k\rho_{k-1}$ and $\rho_k^2$ very near each other for large $n$. Also, as $n \to \infty$,

(19) $$\theta_k \to 0, \quad \theta_k \sim \sin\theta_k$$

In view of (18) and (19) we have, therefore, from (17):

$$\theta = M_o\widehat{M}_n = \sum_{k=1}^{n} M_{k-1}\widehat{M}_k = \sum_{k=1}^{n}\theta_k \sim \sum_{k=1}^{n}\sin\theta_k \sim$$

$$\sim \sum_{k=1}^{n}\frac{t}{n}\frac{1}{\rho_k^2},$$

in other words, we have the limit relation

(20) $$\boxed{\theta = \lim_{n\to\infty}\frac{t}{n}\sum_{k=1}^{n}\frac{1}{1+\left(\frac{tk}{n}\right)^2},} \quad (t = \tan\theta)$$

Since, for $|r| < 1$, one has $\frac{1}{1+r} = 1 - r + r^2 - + \cdots$ we can write, for $|t| < 1$, $\left(\text{ i.e., } |\theta| < \dfrac{\pi}{4}\right)$, and $k = 1, 2, \cdots, n$

$$\frac{t}{n}\frac{1}{1+\left(\dfrac{tk}{n}\right)^2} = \frac{t}{n} - \frac{t^3}{n^3}\cdot k^2 + \frac{t^5}{n^5}\cdot k^4 - + \cdots$$

so that, for the sum occurring in (20), we have

$$\text{(21)} \qquad \frac{t}{n}\sum_{k=1}^{n}\frac{1}{1+\left(\dfrac{tk}{n}\right)^2} = t - \frac{t^3}{n^3}\cdot\sum_{k=1}^{n}k^2 + \frac{t^5}{n^5}\cdot\sum_{k=1}^{n}k^4 - + \cdots$$

Let us put

$$\text{(22)} \qquad \boxed{\sigma_p = \frac{1^{p-1} + 2^{p-1} + \ldots + n^{p-1}}{n^p}} \qquad (p = 1, 2, \cdots)$$

From (20), (21) and (22) we obtain:

$$\theta = \lim_{n\to\infty}\left[t - \sigma_3\cdot t^3 + \sigma_5 t^5 - + \cdots\right]$$

*We now permit ourselves to take term-wise limits*, ie.,

$$\text{(23)} \qquad \theta = t - \left(\lim_{n\to\infty}\sigma_3\right)\cdot t^3 + \left(\lim_{n\to\infty}\sigma_5\right)\cdot t^5 - + \cdots$$

In order to find $\lim\limits_{n\to\infty}\sigma_p$, we prove the following lemma:

**Lemma:** If $\{b_n\}$ is a monotonic *increasing* sequence of *positive* numbers tending to infinity, and

$$\text{(24)} \qquad \lim_{n\to\infty}\frac{a_n - a_{n-1}}{b_n - b_{n-1}} = l,$$

then also

$$\text{(25)} \qquad \lim_{n\to\infty}\frac{a_n}{b_n} = l,$$

Assuming the Lemma, for the moment, we shall find $\lim\limits_{n\to\infty}\sigma_p$

Take, in the lemma,

$$a_n = 1^{p-1} + 2^{p-1} + \cdots + n^{p-1} \text{ and } b_n = n^p$$

Then $0 < b_n \uparrow \infty$, and

$$\frac{a_n - a_{n-1}}{b_n - b_{n-1}} = \frac{n^{p-1}}{n^p - (n-1)^p} = \frac{1}{n}\cdot\frac{1}{1-(1-\frac{1}{n})^p} =$$

$$= \frac{1-r}{(1-r)^p} \text{ with } r = 1 - \frac{1}{n}$$

so that

$$\lim_{n\to\infty}\frac{a_n - a_{n-1}}{b_n - b_{n-1}} = \lim_{r\to 1}\frac{1-r}{(1-r)^p} = \lim_{r\to 1}\frac{1}{1+r+r^2+\cdots+r^{p-1}} = \frac{1}{p}$$

The Lemma now yields $\lim_{n\to\infty}\frac{a_n}{b_n} = \frac{1}{p}$, i.e.,

$$\boxed{\lim_{n\to\infty}\sigma_p = \frac{1}{p}} \tag{26}$$

**Proof of the Lemma:** Put $\alpha_n = a_n - lb_n$; then

$\alpha_n - \alpha_{n-l} = (a_n - lb_n) - (a_{n-1} - lb_{n-1}) = (a_n - a_{n-1} - l(b_n - b_{n-1})$

$$\frac{\alpha_n - \alpha_{n-l}}{b_n - b_{n-1}} = \frac{a_n - a_{n-1}}{b_n - b_{n-1}} - l \longrightarrow 0 \text{ as } n \to \infty \tag{27}$$

(See (24)). Therefore, given $\varepsilon > 0$, there exists $n'$ such that

$$\left|\frac{\alpha_n - \alpha_{n-1}}{b_n - b_{n-1}}\right| < \frac{\varepsilon}{2}, \text{ for } n > n'$$

Since $b_n - b_{n-1} > 0$ always, this becomes

$$|\alpha_n - \alpha_{n-1}| < \frac{\varepsilon}{2}(b_n - b_{n-1}) \text{ for } n = n'+1, n'+2, \cdots \tag{28}$$

Hence, for $n > n'$, we can write

$$\alpha_n - \alpha_n' = (\alpha_n - \alpha_{n-1}) + (\alpha_{n-1} - \alpha_{n-2}) + \cdots + (\alpha_{n'+1} - \alpha_n')$$

whence follows by (28)

$$\begin{aligned}|\alpha_n - \alpha_n'| &< \frac{\varepsilon}{2}\Big[(b_n - b_{n-1}) + (b_{n-1} - b_{n-2}) + \cdots + (b_{n'+1} - b_n')\Big]\\ &= \frac{\varepsilon}{2}(b_n - b_n')\end{aligned}$$

i.e.,

$$\left|\frac{\alpha_n - \alpha_n'}{b_n - b_n'}\right| < \frac{\varepsilon}{2} \text{ for } n > n' \tag{29}$$

We have $\alpha_n = (\alpha_n - \alpha_n') + \alpha_n'$ so that

$$\frac{\alpha_n}{b_n} = \frac{\alpha_n - \alpha_n'}{b_n - b_n'} \cdot \frac{b_n - b_n'}{b_n} + \frac{\alpha_n'}{b_n}$$

Since $0 < b_n \uparrow \infty$, one has here $0 < \frac{b_n - b_n'}{b_n} < 1$, so that

$$\left|\frac{\alpha_n}{b_n}\right| \le \left|\frac{\alpha_n - \alpha_n'}{b_n - b_n'}\right| + \left|\frac{\alpha_n'}{b_n}\right| \tag{30}$$

Here $n'$ is a fixed index, and $\frac{\alpha'_n}{b_n} \to 0$ as $n \to \infty$, so that there exists $n''$ such that

$$\left|\frac{\alpha'_n}{b_n}\right| < \frac{\varepsilon}{2} \text{ for } n > n''$$

From (29), (30) and (31) there follows:

$$\left|\frac{\alpha'_n}{b_n}\right| < \frac{\varepsilon}{2} + \frac{\varepsilon}{2} \text{ for } n > n_0 = \max(n', n'')$$

i.e., $\frac{\alpha'_n}{b_n} \to 0$ as $n \to \infty$. But $\frac{\alpha'_n}{b_n} = \frac{a_n}{b_n} - l$. Consequently

$$\lim_{n\to\infty} \frac{a_n}{b_n} = l$$

which yields (25) and establishes the lemma. ■

From (23) and (26) now follows:

**Theorem. [Mādhava, 14th Century]**

$$(32) \qquad \boxed{\begin{aligned} &\theta = \tan\theta - \frac{\tan^3\theta}{3} + \frac{\tan^5\theta}{5} - + \cdots && \left(|\theta| \le \frac{\pi}{4}\right) \\ &\tan^{-1} t = t - \frac{t^3}{3} + \frac{t^5}{5} - + \cdots && (|t| \le 1) \end{aligned}}$$

We had assumed in the proof that $|\theta| < \frac{\pi}{4}$ (or $|t| < 1$). The series in (32) remain convergent even when $\theta = \pm\frac{\pi}{4}$.

In particular, we have the **Mādhava series for** $\pi$:

$$(33) \qquad \boxed{\frac{\pi}{4} = 1 - \frac{1}{3} + \frac{1}{5} - \frac{1}{7} + - \cdots}$$

The passage to term-wise limits in (23) can be justified, but we shall not do it here.

We have pointed out earlier that the series (32) converges very slowly. In the memoir:

*Srinivasa Ramanujan*
≪Modular equations and approximations to $\pi$≫
$N^0$ 6, Collected Papers, Cambridge, (1927),
[= Quarterly Journal of Mathematics pp 23-39,
XLV (1914) 350-372]

Ramanujan discovered several rapidly converged series for $\pi$:

$$(33) \qquad \frac{4}{\pi} = 1 + \frac{7}{4}\left(\frac{1}{2}\right)^3 + \frac{13}{4^2}\left(\frac{1\cdot 3}{2\cdot 4}\right)^3 + \frac{19}{4^3}\left(\frac{1\cdot 3\cdot 5}{2\cdot 4\cdot 6}\right)^3 + \cdots$$

(34) $$\frac{16}{\pi}=5+\frac{47}{64}\left(\frac{1}{2}\right)^3+\frac{89}{64^2}\left(\frac{1\cdot3}{2\cdot4}\right)^3+\frac{131}{64^3}\left(\frac{1\cdot3\cdot5}{2\cdot4\cdot6}\right)^3+\cdots$$

(35) $$\frac{32}{\pi}=\left(5\sqrt{5}-1\right)+\frac{47\sqrt{5}+29}{64}\left(\frac{1}{2}\right)^3\left(\frac{\sqrt{5}-1}{2}\right)^8+$$

$$+\frac{89\sqrt{5}+59}{64^2}\left(\frac{1\cdot3}{2\cdot4}\right)^3\left(\frac{\sqrt{5}-1}{2}\right)^{16}+\cdots$$

Herein {1, 7, 13, 19, ...} in (33), {5, 4, 7, 89, 131...} in (34), and $\{5\sqrt{5}-1, 47\sqrt{5}+29, 89\sqrt{5}+59,\cdots\}$ in (35) are all in arithmetic progression.

It is said that Hardy was astonished at the originality and profundity of the very first results communicated to him by Ramanujan. Such extraordinary skill and inventiveness in the manipulation of complicated formulas has occurred only a few times in the history of Mathematics: Euler, Gauss, Jacobi had such prowess. Ramanujan had no formal education in Mathematics. He left his proofs lacking rigour. But pioneers and path-finders, exploring boldly new terrains of mathematical thought, permit themselves a freedom in their attack on a mathematical problem. Their intuition often provides them with an innate feeling for what is correct and what is not. It is not a difficult task to edit the works of great savants of by-gone ages so as to satisfy the current demands in mathematical exposition. One cannot expect Madhava in 14th Century to raise and justify the question of term-wise passage to the limit in (23) [See P.187].

Ramanujan's rapidly convergent series for $\pi$ have recently evoked great interest. They have been studied from diverse points of view. [See *Borwein J. and Borwein P.* ≪ Pi and the AGM ≫ (Canadian Math.Soc.) John Wiley (1986)]. In 1987, at the University of lllinios, Urbana-Champaign, USA, there took place an international conference to commemorate the centenary of the birth of Ramanujan (1887-1920) [See ≪ *Ramanujan revisited* ≫. Edited by R. Askey, B. Berndt, K.G. Ramanathan and R.A. Rankin, Academic Press, (1988)]. We wish to quote from this a very striking result of G. Chudnovsky and D. Chudnovsky reported in this conference. In order to state this result we need some notation from the theory of modular functions. Let $\tau$ be a complete variable with imaginary part positive: $Im\ \tau > 0$. Define

$$g_2(\tau) = 60 \sum_{\substack{m,n=-\infty \\ (m,n)\neq(0,0)}}^{\infty} \frac{1}{(m+n\tau)^4},$$

$$g_3(\tau) = 140 \sum_{\substack{m,n=-\infty \\ (m,n)\neq(0,0)}}^{\infty} \frac{1}{(m+n\tau)^6},$$

$\triangle(\tau) = g_2^3(\tau) - 27g_3^2(\tau)$, It is known that $\triangle(\tau) \neq 0$ for all $\tau$.

$$\boxed{j(\tau) = \frac{12^3 g_2^3(\tau)}{\triangle(\tau)}} \qquad (Im\ \tau > 0)$$

Let $\boldsymbol{Q}$ be the field of rational numbers. Consider the number field $\boldsymbol{Q}(\sqrt{-163})$. It is known that it has class number 1. From this information it can be derived that

$$\boxed{\begin{aligned} j(\tau_0) &= \text{ an integer } = -(640320)^3 \\ \text{for } \tau_0 &= \frac{1+\sqrt{-163}}{2} \end{aligned}}$$

Based on these facts, G. Chudnovsky & D. Chudnovsky obtain a very rapidly convergent series for $\pi$:

$$\boxed{\frac{1}{\pi} = 12 \sum_{n=0}^{\infty} (-1)^n \frac{(6n)!}{(n!)^3(3n)!} \frac{13\,59\,1409 + 545140134\,n}{(640\,320)^{3n+\frac{3}{2}}}}$$

This series converges to 15 decimal places per term, and has turned out to be extremely effective for calculation of $\pi$ to a high degree of precision.

## Appendix

The author would like to acknowledge the kind courtesy of Robert Bruce, Lord Balfour of Burleigh for sending a copy of Whish's memoir.

**XXXIII**. *On the Hindù Quadrature of the Circle, and the infinite Series of the proportion of the circumference to the diameter exhibited in the four Śástras, the Tantra Sangraham, Yucti Bháshá, Carana Padhati, and Sadratnamála. By CHARLES M. WHISH, Esq., of the Hon. East-India Company's Civil Service on the Madras Establishment.*

*(Communicated by the MADRAS LITERARY SOCIETY AND AUXILIARY ROYAL ASIATIC SOCIETY.)*

Read the 15th of December 1832.

A'RYAB'HATTA, who flourished in the beginning of the thirty-seventh century of the *Cáli Yuga,** of which four thousand nine hundred and twenty years have passed, has in his work, the *Aryb'hatiyam*, in which he mentions the period of his birth, exhibited the proportion of the diameter to the circumference of the circle as 20000 to 62832, in the following verse:

*Chaturadhicam ṣatamashṭaguṇandwáshashṭistathá sahasranam*
*Ayutadwaya vishcambhasyásannò vritta pariṇáhah.*†

Which is thus translated:

"The product of one hundred increased by four and multiplied by eight, added to sixty and two thousands, is the circumference of a circle whose diameter is twice ten thousand."

The author of the *Lilávati*, who lived six centuries after A'RYAB'HATTA, states the proportion as 7 to 22, which, he adds, is sufficiently exact for common purposes. As a more correct or precise circumference, he proposes that the diameter be multiplied by 3927, and the product divided by 1250; the quotient will be a very precise circumference. This proportion is the same with that of A'RYAB'HATTA, which is less correct than that of METIUS, viz. 113 to *355*, though in all three the circumference is rated at more than the

---

* Or the sixth century of the Christian era.

† This verse is in the variety of the *Aryavrittam* measure, called *Vipula*.

truth. The *slòca* in the *Lilávati* is as follows:

*Vyásè bhanandagnihatè vibhactè*
*Chabáṇasùryaih paridhissussùcshmah*
*Dwávimṣatinighnèvihritètha sailaih*
*Stùlòthavásyádvyavahárayògyah."* *

A'RYAB'HATTA has not left behind him any record of the means by which he obtained so accurate a proportion. His work however shews so much geometrical knowledge, that there can be little doubt that he effected his approximation in the same manner with ARCHIMEDES, by a numeral calculation of the perimeters of inscribed and circumscribed polygons with the circle; for in the polygon of 768 sides to a circle whose radius is *one*, the perimeter of the inscribed polygon is 628316, and of the circumscribed 628326. The circumference of the circle will therefore be greater than 628316, but less than 628326, or nearly half their sum, namely, 628321, to which the above approximation is nearly similar.

In the *Sùrya Sid'dhánta*, the radius of a circle is stated at 3438, in terms of minutes of the circumference, being about one-quarter of a minute too great; and in the following *slòca*, it appears that a very coarse rule is laid down by the author.

*Yojanánám ṣatányashṭau bhùkarnò dwigunánitu*
*Jadvargatò daṣaguṇát padam bhùparidhirbhavèt.*†

"Square the *yojanas* of the earth's diameter 1600, and multiply the square by 10, and extract the root of the product: the root is the equatorial circumference of the earth."

In such a matter, the strictest accuracy need not be required; but for a general rule, this would err considerably from the truth, being less correct than the proportion of 7 to 22; and yet we find the same in the *Cámadògdhri*, a commentary on the *Sùrya Sid'dhánta*, laid down in the following terms:

Vritta vargát daṣa hṛitat padamvyásòbhavèdiha
Evam paricshitòvyásah vṛittastu viparìtatah.

"The root of the square of the circumference divided by ten, is the diameter; to find the circumference of a given diameter, reverse the

---

* This verse is in the *Upajátivrittam* measure.

† This verse is in the *Anushtubvrittam* measure.

rule."

In the *ṡástra* named *Tántra Sangraha*, we find the following lines:

*Ishṭa vyásè hatè nágavèdavahnyabdhichèndubhih*
*Tithyaṣwivibudhairbhactè susùcshmah paridhirbhavèt*
*Paridhèrvyatyayáchaivamsusùcshmam vyásamánayèt*
*Tithyaṣwivibudhairnighnát chachatancènduǹètratah*
*Sùcshmò vyásòhi vèdagni cṛitòbhrèndu hṛitòthava*
*Triṣaghna chacra liptábhyò vyásòrtthèshwagnibhirhṛitèh.*

Which are thus translated:

"Multiply any given diameter by 104348, and divide the product by 33215; the quotient is a very correct circumference. To find the diameter, if the circumference be given, reverse the above numbers. If then you multiply the astronomical circumference by 33215, and divide the product by 104348, you will have the diameter *in terms of minutes of the periphery; or it will be sufficiently exact if you multiply the circumference by* 113, *and divide by* 355, *for a diameter*."

The astronomer's circle, *anantapura* or "the universe," and *chachatancendunetram*, is, according to alphabetical notation and symbols respectively, 21600, that is, the terms of the minutes in the circumference of twelve signs, containing each thirty degrees. In the first proportion of 33215 to 104348, if the diameter of a circle be 1, the circumference will be 3.141592653921, &c., which is an excellent approximation, being correct to the ninth place of decimals, the tenth being too large; the second proportion, 113 to 355, that of METIUS, is a very useful one, being as 1 to 3.1415929. &c., correct to the sixth place of decimals, the seventh being rather more than the truth.

In another work, named *Carana-Padhati*, and in one called *Sadratnamála*, are the two following verses; the first is from the former work:

*Anùnanutnánananunnanutyaih*
*Samáhataṣchacra kálávibhactah*
*Chandámṣuchandrádhama cumbhipálaih*
*Vyásastadardhantribhamaurvicásyát.*

"If the circumference of a circle in minutes be multiplied by 10000000000, and the product be divided by 31415926536, the quotient will be the diameter of the circle *in terms of the minutes of the circumference*, and its half will be the radius."

In the latter work is the subjoined verse:

*Evanchátra parárdha vistṛitimahá vṛittasya nahocsharaih*
*Syádbhadrámbudhi sidha janma ganitasradáṣmayadbhápùgih.*

"If you proceed thus (as laid down in the former verse), and measure the circumference of a great circle by 100000000000000000 parts, the circumference will be equal to 314159265358979324 of *such parts*."

The approximations to the true value of the circumference with a given diameter, exhibited in these three works, are so wonderfully correct, that European mathematicians, who seek for such proportion in the doctrine of fluxions, or in the more tedious continual bisection of an arc, will wonder by what means the *Hindù* has been able to extend the proportion to so great a length. Some quotations which I shall make from these three books, will shew that a system of fluxions peculiar to their authors alone among *Hindùs*, has been followed by them in establishing their quadratures of the circle; and a few more verses, which I shall hereafter treat of and explain, will prove, that by the same mode also, the sines, cosines, &c. are found with the greatest accuracy.

I proceed to quote extracts from the *Tántra Sangraham*. The first, of the measure called *Ánushtubvrittam*, is from the chapter upon sines, &c.

*Vyásárdham prathamannitwatatòványátguṇannayèt*
*Sambandhanniyamanchaivam vynèyòvyása vṛittayòh.*

"Having found the radius, you may construct the sines; but you must first know the proportion between the diameter and the circumference."

The next is of the *Gìtivrittam* measure:

*Vyásè varidhinihitè rùpa hṛitè vyásaságarábhihate*
*Triṣarádhi vishama sanchyá bhactamṛiṇamswam prithacramát curyát*
*Yatsanchyayátra haraṇè critèrnivrittá hṛitistujámitayá*
*Tasyá ùrdhwa gatáyássamasanchyá taddalamgunòntèsyát.*
*Jadvarggò rùpayutò hárò vyásábdhighátacah prágwat.*
*Tabhyámáptam swamṛiṇè critèdhanè ṣòdhananchacaraṇìyám*
*Sùcshmah paridhissasyát bahucritwòharaṇatòti sùcshmascha.*

"Multiply the diameter by 4, and from it subtract and add alternately the quotients obtained by dividing four times the diameter by the odd numbers 3, 5, 7, 9, 11 &c., do thus to the extent required; and having fixed a limit, take half the even number next less than the last odd divisor for a multiplier, and its square *plus* one for a divisor. Multiply four times the diameter by the multiplier, and divide the product by the divisor, and add it or subtract it, according to the sign of the last quote in the series, from the sum of the series, thus the circumference of the given diameter will be obtained very correctly."

If we proceed according to the rule, we have an infinite series of the following form:

$$C = 4d - \frac{4d}{3} + \frac{4d}{5} - \frac{4d}{7} + \frac{4d}{9} - \frac{4d}{11} + \&c. \pm \frac{4d \times \frac{1}{2}p}{p^2+1}$$

where $C$=circumference, $d$ diameter, and $p$ the last odd divisor diminished by unity. When $d$=1 the series becomes

$$C = 4 \times \left(1 - \frac{1}{3} + \frac{1}{5} - \frac{1}{7} + \frac{1}{9} - \frac{1}{11} \pm \&c. \frac{\frac{1}{2}p}{p^2+1}\right)$$

Then follows a verse of the *Gitivrittam* measure, explaining more fully the correction by which this series is brought to greater perfection.

*Asmat sucshmataròny̆ò vilichyatècaschanápi samscárah*
*Antè samasanchyá dala varyassaicò gunassá èva punah*
*Yuga gunitò rùpayuttassamasancyá dala*
*hatò bhavèddhárah*
*Triṣarádi vishama sanchyá haranát paramè tadèva*
*vácáryam.*

"I now shew how the correction may be made more complete than in the former rule: take the even figure next greater than the last odd divisor in the series $4\times\ (1 - \frac{1}{3} + \frac{1}{5}$ - &c.) that may have been fixed upon, and square its half, and *increase* it by *unity*; this is to be a multiplier: this multiplier multiply by 4, and the product increased by unity multiply by half the original even figure; this last product will be a divisor; add to the result of the series the quotient of four times the diameter multiplied by the new multiplier, and divided by the new divisor; the sum will be a more correct circumference."

The series by means of this correction becomes:

$$C = 4 \times \left(1 - \frac{1}{3} + \frac{1}{5} - \frac{1}{7} + \frac{1}{9} - \frac{1}{11} + \&c. \pm \frac{\frac{p^2}{4}+1}{(p^2+4+1)\frac{p}{2}}\right)$$

$p$ being here the last odd divisor increased by unity.

The author being aware how slowly the series converges, found it necessary to correct the last quote, which is done very correctly by the rule he has exhibited.

Next follows a verse of the *Anushtubvrittam* measure.

*Vyásavargádravihatat padamsyát prathamam phalam*
*Jatastattat phaláchchápiyávadişchántribhirharèt*
*Rupádyayugmasancyábhirllabdèshwèshuyathácramam*
*Vishamánám yutètyactè yugmayògè vṛitirbhavèt.*

"Multiply any given diameter squared by 12, and extract the root of the product; this is the first quote; divide the first by 3 for the second quote; this second, and each so obtained quote, divide by 3 continually, place them in order, and divide them in succession by the odd numbers 1, 3, 5, 7, 9, 11 &c. Then add together the first, third, fifth, &c., quotes; and add together also the second, fourth, sixth, &c. quotes; then subtract the sum of the even from that of the odd, and the remainder will be the circumference of the circle, whose diameter was given."

The series laid down in the verse is thus expressed algebraically:

$$C = \sqrt{12} \times \left(1 - \frac{1}{3 \cdot 3} + \frac{1}{5 \cdot 3^2} - \frac{1}{7 \cdot 3^3} + \frac{1}{9 \cdot 3^4} - \frac{1}{11 \cdot 3^5} + \&c.\right)$$

The next verse is of the Gitivrittam measure:

*Samapancháhatayòyá rupádyayujás chaturghna mùlayutáh.*
*Tabhisshòdaşa guņitádvyását pṛithagá hṛitètu vishamayatè.*
*Sama phala yágè tyactè syátishţa vyása sambhavah paridhih.*

"Divide the diameter multiplied by 16 severally by the fifth power of the odd numbers 1, 3, 5, 7, 9, &c., adding to each fifth power four times its root; of the quotes thus obtained subtract the sum of the second, fourth, sixth, &c. from that of the first, third, fifth, and seventh, the remainder will be the circumference of the circle whose diameter was taken."

This is an infinite series of the following form:

$$C = \frac{16d}{1^5 + 4.1} - \frac{16d}{3^5 + 4.3} + \frac{16d}{5^5 + 4.5} - \frac{16d}{7^5 + 4.7} + \frac{16d}{9^5 + 4.9} - \&c.$$

Or, if the diameter by one:

$$C = 16 \times \left(\frac{1}{1^5 + 4.1} - \frac{1}{3^5 + 4.3} + \frac{1}{5^5 + 4.5} - \frac{1}{7^5 + 4.7} + \&c.\right)$$

The next verse of the *Gìtivrittam* measure is as follows:

*Vyásádwaridhi nihatát pṛitagáptantryáyugvimùld ghanaih*
*Trighna vyásè swamriṇam cramasah kritwápi*
*paridhivánèyah.*

"Divide four times the given diameter by the cubes of the odd numbers 3, 5, 7, 9, 11, 13, &c., subtracting from each its respective root; the quotes thus obtained alternately add to, and subtract from three times the diameter; this is the circumference of the circle whose diameter was given."

This is an infinite series (given in the *Carana Padhati* in nearly the same words), as follows:

$$C = 3d + \frac{4d}{3^3 - 3} - \frac{4d}{5^3 - 5} + \frac{4d}{7^3 - 7} - \frac{4d}{9^3 - 9} + \&c.$$

Which, if $d = 1$, will be:

$$C = 3 + 4 \times \left( \frac{1}{2.3.4} - \frac{1}{4.5.6} + \frac{1}{6.7.8} - \frac{1}{8.9.10} + \&c. \right)$$

The author then proceeds in the same measure:

*Dwyádiyujám vacritayò vyècá hárádwinighna vishcambhe*
*Dhanamrinamantèntyòrdhwa gataujacritidwi sahitáharòdwighni.*

"Multiply the diameter by 4, and divide the product severally by the squares of the numbers 2, 4, 6, 8, 10, &c., subtracting 1 from each square: the quotients alternately add to, and subtract from, twice the diameter: rectify the sum obtained, by taking the next odd number less than the last even figure squared, squaring it, adding 2 to the square, doubling this sum, and with this thus obtained as a divisor, dividing four times the diameter: this quotient add or subtract, as is required, from the sum formerly obtained, for a very correct circumference."

The series thus obtained is:

$$C = 2d + \frac{4d}{2^2 - 1} - \frac{4d}{4^2 - 1} + \frac{4d}{6^2 - 1} - \frac{4d}{8^2 - 1} + \&c. \pm \frac{4d}{(p-1)^2 + 2 \times 2}$$

Where $C$= circumference, $d$= diameter, and $p$=the last even number squared in the series, which, if the diameter be 1, becomes,

$$C = 2 + 4 \times \left( \frac{1}{1.3} - \frac{1}{3.5} + \frac{1}{5.7} - \frac{1}{7.9} + \&c. \pm \frac{1}{(p-1)^2 + 2 \times 2} \right)$$

This series, viz. $\frac{1}{1.3} - \frac{1}{3.5} + \frac{1}{5.7} - \frac{1}{7.9} + \&c.$ can easily be proved to be equal to the arc of $90° - \frac{1}{2}$, therefore 2 + 4 into the series will equal the whole circumference, the diameter being *one*, as in our author's series.

Two other series are then exhibited in lines of the *Gìti* measure.

*Dwyádèschaturádèrva chaturadhicánánnirèca vargassyuh*
*Háráh cunjara gunitá vishcambhaswamati calpitòbhájyah*
*Phalayutirádyè vrittam bhájyadalam phalavihìnamányatra.*

"Take the squares of the terms, diminished by 1, of the two arithmetical progressions whose first terms are respectively 2 and 4, and ratio of progression 4, for divisors: in the former series, divide 8 times the diameter by the divisors severally, and the sum of the quotients is the circumference of the circle; in the latter, subtract from 4 times the diameter the sum of the quotients of 8 times the diameter divided by the divisors severally, and the result is the circumference in the second case."

The two series thus explained are found to be of the following forms:

$$C = \frac{8d}{2^2-1} + \frac{8d}{6^2-1} + \frac{8d}{10^2-1} + \frac{8d}{14^2-1} + \frac{8d}{18^2-1} + \&c$$

$$C = 4d - \left(\frac{8d}{4^2-1} + \frac{8d}{8^2-1} + \frac{8d}{12^2-1} + \frac{8d}{16^2-1} + \&c.\right)$$

which, if d = 1, become respectively:

$$C = 8 \times \left(\frac{1}{1.3} + \frac{1}{5.7} + \frac{1}{9.11} + \frac{1}{13.15} + \frac{1}{17.19} + \&c.\right)$$

$$C = 4 - 8 \times \left(\frac{1}{3.5} + \frac{1}{7.9} + \frac{1}{11.13} + \frac{1}{15.17} + \frac{1}{19.21} + \&c.\right)$$

Now, let the former series within the brackets be $a$, and the latter $b$; then it can be easily proved that $a + b$, or the series $\frac{1}{1.3} + \frac{1}{3.5} + \frac{1}{5.7} + \frac{1}{7.9} + \frac{1}{9.11} + \&c. = \frac{1}{2}$; and it has been mentioned above that the series $\frac{1}{1.3} - \frac{1}{3.5} + \frac{1}{5.7} - \frac{1}{7.9} + \&c. =$ (when the diameter is one), the are of $90° - \frac{1}{2} = a - b$; having therefore the values of $a - b$ and $a + b$, it can be easily proved that $a=$ the arc of 45°, and $b = \frac{1}{2} -$ arc of 45°; therefore, in the first of these quadratures, 8a = circumference, and in the second, $4 - 8b$=circumference also, as taught by the author.

The author, after laying down the above series, proceeds to shew in numbers a proportion of the circumference to the diameter in the verse which has been originally quoted and translated, and then finishes his chapter by rules for finding the sines and cosines, of which mention will be made hereafter.

In the *Carana Padhati*, the sixth chapter commences thus:

*Vyásácchaturghnát bahuṣah pṛithac sthát*
*Tripanja saptadyayu gáhritáni*
*Vyasè chatuoghnè cramaṣastwṛiṇam swam*
*Curyáttadásyát paridhissusùcshmah.*

"Divide the given diameter multiplied by 4 severally and continually by the odd numbers 3, 5, 7, 9, 11, &c., and the quotients thus obtained, alternately subtract from, and add to, the diameter multiplied by 4. The result is the precise circumference".

This series is the same with the first laid down by the author of the *Tantra Sangraha*, namely,

$$C = 4 \times \left(1 - \frac{1}{3} + \frac{1}{5} - \frac{1}{7} + \frac{1}{9} - \frac{1}{11} + \&c.\right)$$

The next verse in the chapter is of the measure called *Gitivrittam.*

*Vyásádwana samguṇitát prithagáptantryádyay*
*ugvimùlaghanaih*
*Iriguṇita vyásè swamṛiṇam Cramaṣah*
*cṛitwápi paridhiránèyah.*

"Divide the given diameter multiplied by 4 severally and continually by the cubes of the odd numbers 3, 5, 7, 9, 11, &c., subtracting from each cube its respective root; the sums thus obtained alternately add to, and subtract from, 3 times the diameter: thus you will obtain the circumference of the circle whose diameter was given."

This series is also the same with one of the former book, namely,

$$C = 3 + 4 \times \left(\frac{1}{2 \cdot 3 \cdot 4} - \frac{1}{4 \cdot 5 \cdot 6} + \frac{1}{6 \cdot 7 \cdot 8} - \frac{1}{8 \cdot 9 \cdot 10} + \&c.\right)$$

The fourth verse of the chapter is of the measure called *Indra Vajravrittam.*

*Vargairyujámvadwigunairnnirècaih*
*Vargìcritairvarjitayugma vargaih*
*Vyásancha shadghnam vibhajèt phalam swam*
*Vyásè trinighnè paridhistadásyat.*

"Add to three times the diameter the sum of the quotes obtained by dividing six times the diameter by the square of twice the square minus one of the even numbers 2, 4, 6, 8, 10, &c., subtracting from each the square of its even figure respectively. The sum is the circumference."

This is an infinite series thus expressed algebraically:

$$C = 3d + \frac{6d}{(2 \cdot 2^2 - 1)^2 - 2^2} + \frac{6d}{(2 \cdot 4^2 - 1)^2 - 4^2} + \frac{6d}{(2 \cdot 6^2 - 1)^2 - 6^2} + \&c.$$

Which, if the diameter = 1, becomes

$$C = 3{+}6{\times}\left(\frac{1}{1 \cdot 3 \cdot 3 \cdot 5} + \frac{1}{3 \cdot 5 \cdot 7 \cdot 9} + \frac{1}{5 \cdot 7 \cdot 11 \cdot 13} + \frac{1}{7 \cdot 9 \cdot 15 \cdot 17} + \frac{1}{9 \cdot 11 \cdot 19 \cdot 21} + \&c.\right)$$

The author proceeds with the verse originally quoted, for determining the diameter and radius in terms of minutes of the circumference, and then teaches how, by certain series, the sines, cosines, &c. are to be constructed. He next exhibits a stanza for finding an arc of the circumference of a circle by means of the sum obtained by multiplying the sine of the arc into its radius and dividing the product by its cosine. This sum, it will be readily observed, is equivalent to a tangent of that arc, for cosine is to radius as sine is to tangent; but the tangent individually does not appear in the mathematics of the *Hindùs*.

The lines are as follow:

*Vyásárdhènahatádabhishṭa guṇatah Còtyaptamádyam phalam*
*Jyávar'ena vinighnamádimaphalam tattat phalanchá harèl*
*Crityá còti guṇasya tatratuphalèshwècatripanchá dibhi*
*Bhuctèshwòjayutèstyajèt samayutim jìvádhanussishtyatè.*

"Multiply the given sine by radius, and divide the product by cosine for the first quote; multiply this quote by the square of the sine, and divide the product by the square of the cosine for the second quote: multiply and divide this last quote, and so continually each obtained quote, by the square of the sine and the square of the cosine respectively, and the quotes obtained by this means divide in succession by the odd numbers 1, 3, 5, 7, 9, 11, &c.; then the sum of the 2d, 4th, 6th, 8th, &c. quotes being subtracted from the sum of the 1st, 3d, 5th, 7th, 9th, &c. quotes, the remainder will be the arc of the sine which was taken."

Note—*If the cosine be less than the sine of the given angle, change the names of the two, and proceed as laid down in the rule.*

The infinite series thus beautifully and concisely expressed in the Sanscrit slòca, is the following:

$$a = \frac{r \times s}{\cos.} - \frac{r \times s^3}{3\cos.^3} + \frac{r \times s^5}{5\cos.^5} - \frac{r \times s^7}{7\cos.^7} + \frac{r \times s^9}{9\cos.^9} - \&c.$$

where $a$= arc; $r$= radius; $s$= sine, and cos= cosine.

In this series, the first quote equals the tangent of the arc = t; the second quote equals tangent cubed, divided by thrice radius squared; the third equals tangent to the fifth power divided by 5 into radius to the fourth power; or, algebraically, thus:

$$a = t - \frac{t^3}{3r^2} + \frac{t^5}{5r^4} - \frac{t^7}{7r^6} + \frac{t^9}{9r^8} - \&c.$$

Which is easily proved to be true by a process in fluxions, which demonstrates the fact that the fluxion of the tangent of an arc is to the fluxion of the arc itself as the square of the secant is to the square of the radius; in which case, the fluxion of the arc is proved to be equal to $\frac{r^2\dot{t}}{r^2+t^2}$ and if $r^2\dot{t}$ be divided by $r^2+t^2$, the quotient will be $\dot{t} - \frac{t^2\dot{t}}{r^2} + \frac{t^4\dot{t}}{r^4} - \frac{t^6\dot{t}}{r^6} + \frac{t^8\dot{t}}{r^8} - 7c. = \dot{a}$; and the fluents of each being taken, it becomes $t - \frac{t^3}{3r^2} + \frac{t^5}{5r^4} - \frac{t^7}{tr^6} + \frac{t^9}{9r^8} - \&c. = a =$ the arc itself, as is laid down in the *Carana Padhati*.

It is also seen that from this verse the matter is derived of the former series; for if (in the first series) the arc of 45° be taken, then the tangent will be equal to radius; and if radius = 1, the above series becomes $1 - \frac{1}{3} + \frac{1}{5} - \frac{1}{7} + \frac{1}{9} - \frac{1}{11} + \&c. = a$; and four times this sum will be the semi-circumference, or when the diameter is 1, the whole circumference; therefore $4 \times \left(1 - \frac{1}{3} + \frac{1}{5} - \frac{1}{7} + \frac{1}{9} - \frac{1}{11} + \&c.,\right)$ the circumference, as shown in the *Tantra Sangraha*, and in the *Carana Padhati*.

I proceed now to quote some verses from the *Sadratnamála*. The first which I shall extract is from the chapter on sines, and is of the measure called *Sálini-vrittam*.

*Vargádwyásasyárcanighnat padamyat*
*Tatryamṣò yastècha tattannavámṣáh*

*Dwigna vyècaicadwi purvaujayugmah*
*Chinnányaicya dwyantaram vṛittanáhah.*

"Square the diameter and multiply the product by 12, and extract the root of this product; the root obtained will be the *modulus* of odd quotes, which if you divide by 3, the quotient will be the *modulus* of even quotes. Divide each modulus continually by 9, and the quotient thus obtained from the former, divide by double the numbers 1, 3, 5, 7, 9, &c. minus 1 respectively, and the quotient obtained from the latter, by double the number 2, 4, 6, 8, 10, &c. minus 1 respectively, add up the new obtained quotes, and subtract the sum of those gotten from the even from the sum of those gotten from the odd modulus, the remainder is the circumference of the circle."

The next verse of the *Sarddùla-viccridita* measure is this:

*Vyasárcaghnacriteh padegnibhiraṭònitècha tattat phaláh*
*Cháthaicyadyayugá hṛiteshu paridhirbhèdòyugòjaicyayòh*
*Evanchátra parárdha vistṛiti mahávṛittasya náhácsharaih*
*Syátbhadrámbudhi sidha janma*
*gaṇitasṛadhásmayatbhùpagih.*

"Square the diameter and multiply the product by 12, and extract the root of this product; this root divide continually by 3, and the quotients thus obtained by 1,3, 5, 7, 9, 11, &c., and subtract the sum of the 2d, 4th, 6th, 8th of the last obtained quotes from the sum of the 1st, 3d, 5th, 7th, 9th, &c. If you do thus, and measure the diameter of a great circle by 100000000000000000 equal parts, the circumference will be equal to 314159265358979324 of such parts."

The rule laid down in this verse is exactly similar to that communicated by Dr. HALLEY to the Royal Society of London; and is founded upon the principles laid down in another *slòca* of the *Vasantitilaca-vrittam* measure in the same chapter, as follows:

*Cotihṛita triguṇa báhu vadhè chatasmá*
*Tattat phaláchcha bhujavarga hatáttu cotyáh*
*Crityá cṛitèshucha dharágni sarádi bhacte*
*Shwojaicyatastyajatuyugmayutim dhanustat.*

"The radius into the sine divided by the cosine is the first quote: this multiplied by the square of the sine, and divided by the square of the cosine, is the second quote; this second, and those obtained continually in the same way, multiply and divide by the square of the sine and square of the cosine respectively: divide the quotes in order by 1, 3, 5, 7, 9, 11, &c. respectively, and the difference of the sum of the 1st, 3d, 5th, &c., and of the 2d, 4th, 6th, &c., will be the are whose sine was taken."

This, of which the meaning is the same with that in the *Carana Padhati*, is expressed thus:

$$a = \frac{r \times s}{\cos.} - \frac{r \times s^2}{3 \cos.^3} + \frac{r \times s^7}{7 \cos.^7} + \&c.$$

Now if the arc of 30° be taken, the first quote = the tangent of that arc = $\sqrt{\frac{1}{3}}$; to multiply this by the square of the sine, and divide it by the square of the cosine, it is sufficient to divide it simply by 3 (because the square of the sine of 30° is one-third of the square of the cosine of that angle): divide then each quote in succession by the odd numbers 1, 3, 5, 7, 9, 11, &c., respectively; the arc, thus obtained, multiplied by 6, will equal the semi-circumference when the radius = 1, or the whole circumference if the diameter be 1; but $6 \times \sqrt{\frac{1}{3}} = 2 \times \sqrt{3} = \sqrt{12}$; therefore $\sqrt{12} \times \left(1 - \frac{1}{3.3} + \frac{1}{5.3^2} - \frac{1}{7.3^3} = \frac{1}{9.3^4} - \frac{1}{11.3^5} + \&c.\right)$ equals the whole circumference when the diameter is *one*, according to the rule laid down in the *Sadratnamálah* in the verse *"Vyasárcaghnacriteh"*, &c., and in the *Tantra Sangraha* in the verse *"Vyasa Vargadravihatat,"* &c. before quoted. the rule also for finding the sum of any arc from the tangent is thus expressed in the same work in four *slòcas* of the *Anushtub* measure.

*Vrittánayana márginábreshta jyáná dhanurnayèt*
*Ishta jyá trijyayòrghátát còtyáptam prathamam phalam*
*Jyávargom gunacam critwá coti varganchá háracam*
*Prathamádi phalèbhyòthá nèyáphalalatirmuhuh.*
*Ecatryádyája sanchyábhirbhactèshwètèshwanucramát*
*Ojánám yutitastyacte yugma yògè dhanurbhavèt*
*Dòh còtyòralpamèrèshtam calpaniyamihòditam*
*Muhuh critèpi labd' hinánnanyathávasitirbhavet.*

"According to the genius of the rules which we have laid down for the circumference of the circle, the arc of any given sine may be found thus: take a given sine multiplied by radius, and divided by the cosine to the given sine for the first quote; let the square of the sine be a constant multiplier, and the square of the cosine a constant divisor, the quotes obtained by multiplying and dividing (by these) the first quote, must be placed in order one under the other, and divided in succession by the odd numbers 1, 3, 5, 7, 9, 11, &c.; then subtract the sum of the even from that of the odd quotes, and the remainder will be the arc whose sine was taken. If the sine of the given are be greater than its cosine, call that the cosine, and the cosine to it the sine."

Having thus submitted to the inspection of the curious eight different infinite series, extracted from *Bráhmanical* works for the quadrature of the circle, it will be proper to explain by what steps the *Hindù* mathematician has been led to these forms, which have only been made known to Europeans through the method of fluxions, the invention of the illustrious NEWTON. Let us first, however, know the age of these works; and as far as can be determined, the authors. First, then, it is a fact which I have ascertained beyond a doubt, that the invention of infinite series of these forms has originated in Malabar, and is not, even to this day, known to the eastward of the range of *Gháts* which divides that country, called in the earliest times *Cèralam*, from the countries of *Madura, Coimbatore, Mysore*, and those in succession, to that northward of these provinces.

The author of *Sadratnamálah* in SANCARA VARMA, the younger brother of the present *Rájá* of *Cadattanáda* near *Tellicherry*, a very intelligent man and acute mathematician. This work, which is a complete system of *Hindù* astronomy, is comprehended in two hundred and eleven verses of different measures, and abounds with fluxional forms and series, to be found in no work of foreign or other Indian countries.

The author of the *Carana Padhati*, whose grandson in now alive in his seventieth year, was PATHUMANA SÒMA YAJI, a *Nambùtiri Bráhmana* of *Tirusìvapura* Trichùr in Malabar. In the first verse of the commentary on his work, he is thus mentioned:

*Nutuna griha sama sutarachitayah carana padhatervidusha*
*Bhasham vilichati caschit balanam bhodhanarthamalpadhiyam.*

"The *Carana Padhati*, a path to arithmetic, composed by NUTUNA-GRIHA SÒMA SUTA, I now proceed to explain in the common tongue, to adapt it to the genius of young scholars".

NUTUNAGRIHA SÒMA SÙTA bears the meaning in *Sanscrita* of the Malabar terms "PUTHAMANA SÒMA YAJI," the first word being the name of his house.

The last verse of his work contains its date in numerical letters in terms of days of the *Caliyuga*; a mode of dating which, in the country, is to be traced back to upwards of one thousand seven hundred years ago, as may be seen in the inscriptions taken from the pagoda of *Tiruvunnùr* near Calicut. The verse is as follows:

*Iti sivapura nam agramajah Capiyajwa*
*Cimapi Carana padhatyahwayam tantra rupam*
*Vyadhita ganita metutsamyagalocyasantah*
*Cathitamihavidantassuntu santosha pantah.*

The words *Ganita Metutsamyac*, written in numbers, amount to 1765653 in terms of days of the present age; which is found to agree with A.D. 1733, being 86 years and 271 days before the 31st of December 18191 *SÒMA YAJI* was not, however, inventor of the system by which he formed his infinite series.

The author of the *Tantra Sangraha*, who was educated in the college of *Bráhmanas* at Trichùr (which college still exists, and which is a place of reputed sanctity, and certainly of respectable learning), has, among his works on arithmetic and astronomy, laid the foundation for a complete system of fluxions, and opened a mine of wealth in mathematics to those students who can boast of being his followers, of which they are most eminently jealous. The copy of the work, which I have obtained with some difficulty, by frequent intercourse with this interesting society, bears in itself marks of antiquity, the commentary in the vulgar tongue being written in a language which is not now generally current in Malabar, and the forms of many letters differing materially from those of the present day.

The author of the work is *TALACULATTÙRA NAMBÙTIRI*, of Cèrala, or Malabar; the first epithet being the name of his house, the latter the title of his rank: he flourished in the forty-sixth century of the present age, or *Caliyuga*.

The testimonies as to the author, and the period in which he lived, are the following, viz. The general consent of the learned in Malabar; the date which is shown in the commencement of the work itself, namely, the year 4600 of the *Caliyuga*; the mention made of him in the first chapter of a work named *Driccaranam* by his commentator, the author of the *Yucti-Bháshá*, CELLALURA NAMBỤTIRI, in which, while relating the history of the progress of astronomy, from the improvements of A'RYAB'HATTA up to his own period, he thus observes: "Much difference having occurred in astronomical calculations, in the year of *Caliyuga* 4532 a *Bráhman* of high rank who lived on the coast of the western ocean,, having examined the heavens for twelve years, established what is laid down in the *Tantra-Sangraha*". This is the evidence of the author of the *Yucti-Bháshá*, the commentary on the *Tantra-Sangraha*, concerning the author of the latter work: the date of the *Driccaranam* is mentioned in the latter part of the work, viz. the 783d of the Malabar era; and in the summary account of the periods of astronomy, it is written 4708 of the *Caliyuga*, both of which coincide with the year 1608 of the Christian era.

A farther account of the *Yucti-Bháshá*, the demonstrations of the rules for the quadrature of the circle by infinite series, with the series for the sines, cosines, and their demonstrations, will be given in a separate paper: i shall therefore conclude this, by submitting a simple and curious proof of the 47th proposition of EUCLID, extracted from the *Yucti-Bháshá*. In the accompanying figure, let ABC be a triangle, having the angle at C and right angle: on AC describe the square ACFN, and on BC describe the square BCDE: on ED take EG equal to AC, and on AD take AH equal to BC, draw HK perpendicular to AD and equal to AC, join BG, GK, and AK.

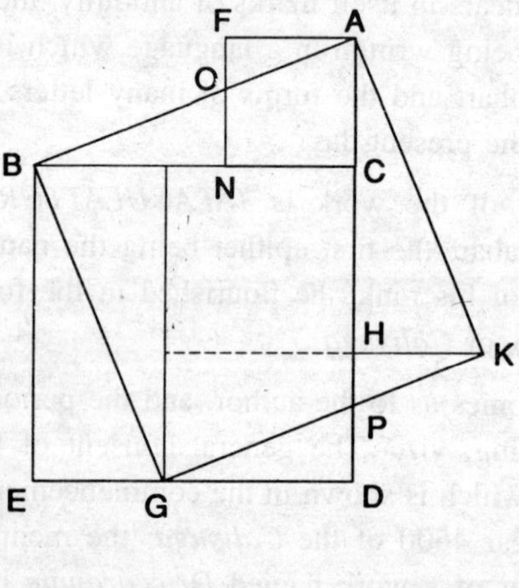


The mathematician will easily prove that the three spaces AFO,

PGD, and BEG, parts of the squares SACFN and BCDE not included in the figure ABGK, are equal to, and identical with, the spaces KHP, OBN, and AHK, not occupied by any parts of the above two squares; *id est*, that the sum of the two squares ACFN and BCDE equals the figure ABGK; but this, from its construction, is a square, and is drawn upon the hypothenuse ABC. This is probably the form by which PYTHAGORAS discovered the celebrated problem, which EUCLID afterwards so beautifully illustrated in the 47th proposition of the first book of his 'Elements.'

SUPPLEMENT–I

# BKT ALGORITHMS

## § MULTIPLICATION AND DIVISION

We present here the Bharathi Krishna Thirtha Division Algorithm.

In the Chapters IV, V and VI of his book the author discusses numbers special examples, and devotes the Chapter XXVII entitled straight division describes the general algorithm with great enthusiasm. We designate his method as the BKJ Division **Algorithm.**

**Notation**

$N$ the set of natural numbers

1, 2, 3,...

For two $n$-tuples in $N$

$x_1,..., x_n$ and $y_1 ..., y_n$

We put

$$(1)\quad \begin{pmatrix} x_1 \dots x_n \\ y_1 \dots y_n \end{pmatrix} = \Sigma x_i - x_j,\ \ (itj = n,\ 0 \leq a, ; \leqq n)x$$

$$= x_1 y_n + x_2 y_{n-1} + \dots + x_n y_1$$

We call it the Skew-product of the two $n$-tuples.

We are particularly interested in the case where the $x_i$ and $y_j$ are restricted to lie in the interval [0, 9], i.e.

$$0 \leqq x_i, y_j \leqq 9$$

Then every natural number $A$ can be written as

$$(2)\quad A = a_0 \cdot 10^n + a_1 - 10^{n-1} + \dots + a_{n-1} \cdot 10 + a_n$$

$$(3)\quad 1 \leqq a_0 \leqq 9,\ 0 \leqq a_i \leqq 9,\ n = 1, \dots, n$$

We call $a_0$, ..., the digits of the number $A$, and $a_0$ the leading digit of $A$. The digits are restricted by the condition (3)

Before we proceed further, we shall look at the following special case which reveals all the features of the general case.

**Example** $D = d_0 d_1 d_2, \quad Q = q_0, q_1, q_2, q_3, q_4, q_5, q_6$

$$\begin{array}{ccccccccccccc}
d_0q_0 & & d_0q_1 & & d_0q_2 & & d_0q_3 & & d_0q_4 & & d_0q_5 & & d_0q_6 \\
& / & & / & & / & & / & & / & & / & \\
d_1q_0 & & d_1q_1 & & d_1q_2 & & d_1q_3 & & d_1q_4 & & d_1q_5 & & d_1q_6 \\
& / & & / & & / & & / & & / & & / & \\
d_2q_0 & & d_2q_1 & & d_2q_2 & & d_2q_3 & & d_2q_4 & & d_2q_5 & & d_2q_6
\end{array}$$

From the diagram it is clear that

$$\sum_{\mu=0}^{2}\sum_{\upsilon=0}^{6} d_\mu \, q_\upsilon = \sum_{k=0}^{8} p_k$$

Where the skew-products $p_k$; ($k = 0, \ldots 8$) are given as follows:

$$p_0 = \begin{pmatrix} d_0 \\ q_0 \end{pmatrix} p_1 = \begin{pmatrix} d_0 d_1 \\ q_0 q_n \end{pmatrix}, \; p_{2+k} = \begin{pmatrix} d_0 d_1 d_2 \\ q_k q_{k+1} q_{k+2} \end{pmatrix}, (k = u_1 \ldots 4),$$

$$p_7 = \begin{pmatrix} d_1 d_2 \\ q_5 q_6 \end{pmatrix} p_8 = \begin{pmatrix} d_2 \\ q_8 \end{pmatrix}$$

Lemma (Diagonal Summation)

$$(4) \qquad \boxed{\sum_{\mu=0}^{j}\sum_{\gamma=0}^{h} d_\mu q_\gamma = \sum_{k=0}^{j+h} p_k, \quad (h \geq j)}$$

where

$$p_k = \begin{pmatrix} d_0 \ldots d_p \\ q_0 \ldots q_k \end{pmatrix}, \quad p_{j+k} = \begin{pmatrix} d_0 \ldots d_j \\ q_k \ldots q_{j+k} \end{pmatrix}, \quad p_{h+k} = \begin{pmatrix} d_0 \ldots d_j \\ q_{h-j+k} \ldots q_n \end{pmatrix},$$

$$(0 \leq k \leq j-1) \qquad (0 \leq k \leq h-j) \qquad (1 \leq k \leq j)$$

**Proof:** the identity (4) is already clear from the special case we have illustrated ($j = 2$, $q = 6$). The sum on the left site of (6) can be broken up into three partial sums.

$$(1) \sum_{k=0}^{j-1} \sum_{\mu+\gamma=j}, \quad (2) \sum_{k=0}^{j} \sum_{\mu+\gamma=j+k}, \quad (3) \sum_{k=1}^{j} \sum_{\mu+\gamma=h+k}$$

i.e., (1) $\sum_{k=0}^{j-1} p_k$, (2) $\sum_{k=0}^{h-j} p_{j+k}$, (3) $\sum_{k=1}^{j} q_{h+k}$

This proves the identity.

The Lemma on diagonal summation yields us at once the following algebraic identity.

$$(6) \quad \boxed{\sum_{\mu=0}^{j} d_\mu x^{j-\mu} x \cdot \sum_{\gamma=0}^{h} q_\upsilon x^{h-\upsilon} \cdot = \sum_{k=0}^{n} p_k x^{n-k}}$$

$(n = j + h, h \geqq j)$.

For our present purpose we need this identify in the special case.

$$(7) \quad \boxed{x = 10,\ d_0 \neq 0,\ q_0 \neq 0,\ 0 \leq d_\mu, q_{\nu\upsilon} \leqq 9}$$

Under these restrictions where

$$\sum_{\mu=0}^{j} d_\mu 10^{i-\mu} = d_0 \ldots d_j, \quad \sum_{\nu=0}^{h} d^\upsilon 10^{h-\upsilon} = q_0 - q_h$$

Consequently we obtain the following,

Theorem (Product Formula) $(n = j + h, h \geq j)$

$$\boxed{(d_0 \ldots d_j)(q_0 \ldots q_h) = \sum_{k=0}^{n} 10^{n-k} \cdot p_k}$$

The BKT Algorithms for multiplication and division rest on this product formula.

**BKT-Multiplication Algorithm**

Given $D = d_0\, d_n \ldots d_j$ and $Q = q_0 q_n \cdots q_n$ $(h \geq j)$ to determine the product $A = a_0\, a_1 \ldots a_n$, $(n = j + h)$

Let $p_0, p_1, \ldots p_n$ $(n = j + h)$ be the skew – products given by $D$ and $Q$ (See Lemma on Diagonal Summation)

$$\left(r_{o=0}\right) \begin{array}{c|ccccccc|} p_k & p_0 & p_n & \cdots & p_k & p_{k+1} & \cdots & p_{n-1} & p_n \\ r_k & & r_1 & \cdots & r_k & r_{k+1} & \cdots & r_{n-1} & r_n \\ v_k & a_0 & a_r & \cdots & a_k & a_{k+1} & \cdots & a_{n-1} & a_n \end{array} \begin{array}{l} \left(r_{n+1} = 0\right) \\ \left(r_k a_k = 10 r_k + a_k\right) \end{array}$$

Starting from the end, $p_n = \begin{pmatrix} d_j \\ q_h \end{pmatrix}$, we divide $p_n$ by 10 to write

$p_n = 10\, r_n + a_n$. Next we divide $p_{n-1} + r_n$ by 10 to obtain $p_{n-1} + r_n = 10r_{n-1} + a_{n-1}$, ...In this way we get the sequence of residues $r_1,... r_n$ and the sequence of digits $a ...a_n$ such that

$$a_0 = p_0 + r_1, ...r_1 a_1 = p_1 + r_2,..., r_n a_n = p_n$$

If we define $r_0 = 0$, and $r_{n+1} = 0$ we have, therefore,

$$\boxed{r_k\, a_k = p_k + r_{k+1}} \qquad (k = 0, 1, ..., n)$$

Hence
$$p_k - q_k = 10\, r_k - r_{k+1},$$
$$10^{n-k}(p_k - a_k) = 10^{n-k+1} r_k - 10^{n-k} r_{k+1}$$

or, putting $t_k = 10^{n-k+1}\, r_k$, we obtain

$$\sum_{k=0}^{n} 10^{n-k}\left(p_k - a_k\right) = \sum_{k=0}^{n}\left(t_k - t_{k+1}\right) = t_0 - t_n = 0,$$

since $r_0 = 0 = r_{n+1}$. By the product formula

$$\sum_{k=0}^{n} 10^{n-k}\, p_k = \left(d_0 ... d_j\right) \times \left(q_0 ... q_h\right)$$

while $\sum_{k=0}^{n} 10^{n-k}\, a_k = a_0\, a_1,\ ...,\ a_n$. Consequently

$$(d_0\, ...d_j) \times (q_0 ...q_h) = a_0 ...a_n$$

**BKT – Division Algorithm**

Given the numbers

$$A = a_0 a_1 ...a_n, \qquad D = d_0 d_n ...d_j$$
$$a_0 \geq d_0,\ n - j = h$$

We determine the quotient $Q$ of $A$ by $D$ as following: $Q = q_0 q_1 ...q_h$. Divide $a_0$ by the leading digit $d_0$ of the divisor $D$ to get

$$a_0 = \begin{pmatrix} d_0 \\ q_0 \end{pmatrix} + r_n$$

Then divide $r_1 a_n - \begin{pmatrix} d_0 \\ q_1 \end{pmatrix}$ by $d_0$ to get

$$r_1 a_n - \begin{pmatrix} d_1 \\ q_0 \end{pmatrix} = \begin{pmatrix} d_0 \\ q_1 \end{pmatrix} + r_2, \text{ i.e., } r_1 a_1 = \begin{pmatrix} d_0 d_1 \\ q_0 q_1 \end{pmatrix} + r_2$$

Continuing in this way we obtain

$$\boxed{r_k a_k = p_k + r_{k+1}} \qquad (k = 0, 1, ...h)$$

The division by the leading digit $d_0$ of the divisor $D$ stops here, i.e. as soon as $q_h$ is obtained. Then

$$r_{h+k}a_{h+k} = p_{h+k} + r_{h+k+1} \qquad (k = 1, ..., j)$$

Consequently, we have

$$\boxed{r_k a_k = p_k + r_{k+1}} \qquad (k = 0, 1, ...n)$$

The skew products are displayed in the Diagonalisation Lemma

$$p_k\begin{pmatrix} d_0 ... d_k \\ q_0 ... q_k \end{pmatrix}(0 \le k \le j-1)\, p_{j+k} = \begin{pmatrix} d_0 ... d_j \\ q_k ... q_{j+k} \end{pmatrix}(0 \le k \le h-j)$$

and $\quad p_{h+k} = \begin{pmatrix} d_{kl} ... d_j \\ q_{h-j+k} ... q_h \end{pmatrix}, \ (1 \le k \le j).$

We now have

$$10\, r_k + a_k = p_k + r_{k+1},\ a_k - p_k = r_{k+1} - 10 r_k,$$

$$10^{n-k}(a_k - p_k) = 10^{n-k} r_{k+1} - 10^{n-k+1} r_k.$$

If – we put $t_k = 10^{n-k} r_{k+1}$, we obtain

$$\sum_{k=0}^{n} 10^{n-k}(a_k - p_k) = \sum_{k=0}^{n}(t_k - t_{k-1}) = t_n - t_0 = r_{n+1}$$

Here $\quad \sum_{k=0}^{n} 10^{n-k} a_k = a_0 ... a_n$, and by the product

formula $\sum_{k=0}^{n} 10^{n-k} p_k = (d_0 ... d_j)(q_0 ... q_h)$, *i.e.*

$$\boxed{a_0 ... a_n = (d_0 ... d_j) \times (q_0 ... q_h) + r_{n+1}}$$

If we put $r_{n+1} = R$,

$$a_0 a_1 ... a_n = (d_0 ... d_j) + (q_0 ... q_h) + r_{n+1}$$

i.e., $\quad \boxed{A = DQ + R}$, $Q = q_0 ... q_b$.

$$R = r_{n+1}$$

**Remarks**

1. We have assumed that $a_0 \geqq d_0$, i.e. the leading digit $d_0$ of the divisor $D$ is less than or equal to the leading digit of the number $A$. This is not a real restriction, for in case $a_0 < d_0$ we shall write
$$A = a_0 a_1 \cdot 10^{n-1} + ...$$
or, we will have to write $a_0 = \begin{pmatrix} q_0 \\ d_0 \end{pmatrix} + a_0$, *i.e.*, $q_0 = 0$, $r_1 = a_0$.

2. We have assumed $h = \geqq j$. However, in the relation $A = DQ + R$, $D$ and $Q$ can be inter-changed, i.e. if one is the divisor the other is the quotient. In other words, the case $j > h$ can be deduced from the case already considered by interchanging $d_\mu$ with $q_\upsilon$.

## SCHOLIUM

### (a) BKI – Multiplication Algorithm

| $d_0\, d_1 ... d_j$ | $p_0\, p_1 \cdots$ | $p_k\, p_{k+1} \cdots$ | $p_{n-1}$ | $p_n$ | |
|---|---|---|---|---|---|
| $q_0\, q_1 ... q_j ... q_h$ | $(r_0 = 0) r_1$ | $r_k\, r_{k+1 ...}$ | $r_{n-1}$ | $r_n$ | $r_{n+1} = 0$ |
| $r_k\, a_k = 10 r_k + a_k$ | $a_0\, a_n$ | $a_k \cdots$ | $a_{n-1}$ | $a_n$ | |

$$a_n = r_n p_n, p_{n-1} + r_m = r_{n-1} + a_{n-1}, ..., r_1 + p_0 = a_0$$

$$\boxed{r_k a_k = p_k + r_{k+1}} \qquad (k = 0, 1, ... n)$$

$$\boxed{d_0 ... d_j \times q_0 ... q_h = a_0 a_1 ... a_n}$$

### (b) BKT – Division Algorithm ($h = n - j$, $h \geqq j$)

| $d_n - d_j$ | $a_0\, a_1\, ...\, a_j$ | $a_{k+1} \cdots a_h$ | $a_{b+1} \cdots q_{h+j}$ | |
|---|---|---|---|---|
| $(d_0)$ | $r_1 \quad ...\, r_j$ | $r_{j+1} \cdots r_h$ | $r_{b+1} \cdots r_{b+j}$ | |
| $(r_0 = 0)$ | $q_0\, q_1\, ...\, q_j$ | $q_{j+1} \cdots q_h$ | | $r_{n+1} = R$ |

$$\boxed{a_0\, a_1 ... a_n = (d_0 ... d_j) \times (q_0 ... q_h) + R}$$

**Example 1.** 6 7 8

| 1 4 5 6 7 1 7 | 6 | 31 | 66 | 103 | 124 | 103 | 105 | 57 | 56 |
|---|---|---|---|---|---|---|---|---|---|
| | | 3 | 7 | 11 | 13 | 11 | 11 | 6 | 5 |
| | 9 | 8 | 7 | 6 | 5 | 4 | 1 | 2 | 6 |

$$p_0=\begin{pmatrix}6\\1\end{pmatrix}=6,\qquad p_1=\begin{pmatrix}6&7\\1&4\end{pmatrix}=31,\qquad p_2=\begin{pmatrix}6&7&8\\1&4&5\end{pmatrix}=66,$$

$$p_3=\begin{pmatrix}6&7&8\\4&5&6\end{pmatrix}=103,\quad p_4=\begin{pmatrix}6&7&8\\5&6&7\end{pmatrix}=124,$$

$$p_5=\begin{pmatrix}6&7&8\\6&7&1\end{pmatrix}=103,\quad p_6=\begin{pmatrix}6&7&8\\7&1&7\end{pmatrix}=105,$$

$$p_7=\begin{pmatrix}7&8\\1&7\end{pmatrix}=57,\qquad p_8=\begin{pmatrix}8\\7\end{pmatrix}=56.$$

**Example 2.**

| | | | | | | | | | | | |
|---|---|---|---|---|---|---|---|---|---|---|---|
| 7 | 8 | 9 | 8 | 7 | 6 | 5 | 4 | 3 | 2 | 1 | |
| | (6) | | 3 | 7 | 11 | 13 | 11 | 11 | 8 | 2 | |
| | | 1 | 4 | 5 | 6 | 7 | 1 | 7 | | | 1 9 5 |

$$9=\begin{pmatrix}6\\1\end{pmatrix}+3,\ 39-\begin{pmatrix}7\\1\end{pmatrix}=\begin{pmatrix}6\\4\end{pmatrix}+7,\qquad 77-\begin{pmatrix}7&8\\1&4\end{pmatrix}=\begin{pmatrix}6\\5\end{pmatrix}+11,$$

$$116-\begin{pmatrix}7&8\\4&5\end{pmatrix}=\begin{pmatrix}6\\6\end{pmatrix}+13,\qquad 135-\begin{pmatrix}7&8\\5&6\end{pmatrix}=\begin{pmatrix}6\\1\end{pmatrix}+11,$$

$$113-\begin{pmatrix}7&8\\1&7\end{pmatrix}=\begin{pmatrix}6\\7\end{pmatrix}+8,\qquad 82-\begin{pmatrix}7&8\\4&7\end{pmatrix}=25,$$

$$251-\begin{pmatrix}8\\7\end{pmatrix}=195.$$

**Example 3.**

| | | | | | | | | |
|---|---|---|---|---|---|---|---|---|
| 2 4 6 8 | 18 | 50 | 92 | 140 | 100 | 62 | 30 | 8 |
| 9 7 5 3 1 | | 6 | 10 | 15 | 10 | 6 | 3 | 0 |
| | 24 | 0 | 7 | 0 | 6 | 5 | 0 | 8 |

$$p_0=\begin{pmatrix}2\\9\end{pmatrix}=18,\qquad p_1=\begin{pmatrix}2&4\\9&7\end{pmatrix}=50,$$

$$p_2 = \begin{pmatrix} 2 & 4 & 6 \\ 9 & 7 & 5 \end{pmatrix} = 92, \qquad p_3 = \begin{pmatrix} 2 & 4 & 6 & 8 \\ 9 & 7 & 5 & 3 \end{pmatrix} = 140,$$

$$p_4 = \begin{pmatrix} 2 & 4 & 6 & 8 \\ 7 & 5 & 3 & 1 \end{pmatrix} = 100, \qquad p_5 = \begin{pmatrix} 4 & 6 & 8 \\ 5 & 3 & 1 \end{pmatrix} = 62,$$

$$p_6 = \begin{pmatrix} 6 & 8 \\ 3 & 1 \end{pmatrix} = 30, \qquad p_7 = \begin{pmatrix} 8 \\ 1 \end{pmatrix} = 8$$

**Example 4.**

| 7 5 3 1 | 24 0 7 0 | 6 5 0 8 | |
|---|---|---|---|
| (9) | 6 10 15 | 10 6 3 0 | |
| | 2 4 6 8 | | 0 |

$$24 = \begin{pmatrix} 9 \\ 2 \end{pmatrix} + 6, \quad 60 - \begin{pmatrix} 7 \\ 2 \end{pmatrix} = \begin{pmatrix} 9 \\ 4 \end{pmatrix} + 10, \quad 107 - \begin{pmatrix} 7 & 5 \\ 2 & 4 \end{pmatrix} = \begin{pmatrix} 9 \\ 4 \end{pmatrix} + 15,$$

$$150 - \begin{pmatrix} 7 & 5 & 3 \\ 2 & 4 & 6 \end{pmatrix} = \begin{pmatrix} 9 \\ 8 \end{pmatrix} + 10, \qquad 106 - \begin{pmatrix} 7 & 5 & 3 & 1 \\ 2 & 4 & 6 & 8 \end{pmatrix} = 6,$$

$$65 - \begin{pmatrix} 5 & 3 & 1 \\ 4 & 6 & 8 \end{pmatrix} = 3, \qquad 30 - \begin{pmatrix} 3 & 1 \\ 6 & 8 \end{pmatrix} = 0, \qquad 08 - \begin{pmatrix} 1 \\ 8 \end{pmatrix} = 0.$$

**Example 5.**

| 1 2 1 3 | 6 | 14 | 10 | 28 | 30 | 24 | 32 | 24 | $p_k$ |
|---|---|---|---|---|---|---|---|---|---|
| 6 2 0 8 8 | | 1 | 1 | 3 | 3 | 2 | 3 | 2 | $r_k$ |
| | 7 | 5 | 3 | 1 | 2 | 7 | 4 | 4 | $a_k$ |

$$p_0 = \begin{pmatrix} 1 \\ 6 \end{pmatrix} = 6, \qquad p_1 = \begin{pmatrix} 1 & 2 \\ 6 & 2 \end{pmatrix} = 14,$$

$$p_2 = \begin{pmatrix} 1 & 2 & 1 \\ 6 & 2 & 0 \end{pmatrix} = 10, \qquad p_4 = \begin{pmatrix} 1 & 2 & 1 & 3 \\ 6 & 2 & 0 & 8 \end{pmatrix} = 28,$$

$$p_5 = \begin{pmatrix} 1 & 2 & 1 & 3 \\ 2 & 0 & 8 & 8 \end{pmatrix} = 30, \qquad p_6 = \begin{pmatrix} 2 & 1 & 3 \\ 0 & 8 & 8 \end{pmatrix} = 24,$$

$$p_7 = \begin{pmatrix} 1 & 3 \\ 8 & 8 \end{pmatrix} = 32, \qquad \begin{pmatrix} 3 \\ 8 \end{pmatrix} = 24$$

**Example 6.**

$$\begin{array}{ccc|ccccc|ccc|ccc}
2 & 1 & 3 & 7 & 5 & 3 & 1 & 3 & 5 & 7 & 9 & & & \\
 & (1) & & & 1 & 1 & 3 & 3 & 3 & 11 & 85 & & & \\
\hline
 & & & 6 & 2 & 0 & 8 & 8 & & & & 8 & 3 & 5
\end{array}$$

$$7 = \begin{pmatrix} 1 \\ 6 \end{pmatrix} + 1, \qquad 15 - \begin{pmatrix} 2 \\ 6 \end{pmatrix} = \begin{pmatrix} 1 \\ 2 \end{pmatrix} + 1, \qquad 13 - \begin{pmatrix} 2 & 1 \\ 6 & 2 \end{pmatrix} = \begin{pmatrix} 1 \\ 0 \end{pmatrix} + 3,$$

$$31 - \begin{pmatrix} 2 & 1 & 3 \\ 6 & 2 & 0 \end{pmatrix} = \begin{pmatrix} 1 \\ 8 \end{pmatrix} + 3, \qquad 35 - \begin{pmatrix} 2 & 1 & 3 \\ 1 & 0 & 8 \end{pmatrix} = \begin{pmatrix} 1 \\ 8 \end{pmatrix} + 3,$$

$$35 - \begin{pmatrix} 2 & 1 & 3 \\ 0 & 8 & 8 \end{pmatrix} = 11, \quad 117 - \begin{pmatrix} 1 & 3 \\ 8 & 8 \end{pmatrix} = 85, \quad 859 - \begin{pmatrix} 3 \\ 8 \end{pmatrix} = 835.$$

SUPPLEMENT–II

# SOME REMARKS ON PELL'S EQUATION AND BHASKARAS CHAKRAVALA METHOD

## § 1. INTRODUCTION

**Notation**

$/N$ is the set of natural number $i$ for $a$, $b$ in $/N$ the greatest common divisor of $a$ and $b$ (g.c.d) is denoted by $(a, b)$ $Q$ is the set of rational numbers; $D$ is in $/N$ but $\sqrt{D}\,x + y$ with $x$, $y$ in $q$; the conjugate $-\sqrt{D}\,x + y$ is denoted by $\bar{z}$, $F(z) = -z\bar{z} = Dx^2 - y^2$; $/N\,[\gamma D]$ the set of numbers $z = \sqrt{D}\,x + y$ with $x$, $y$ in $/N$; $/N_0\left[\sqrt{D}\right]$ is the subset of $/N\left[\sqrt{D}\right]$ consisting of $\sqrt{D}\,x + y$ with $x$, $y$ in $/N$ *and* $(x, y) = 1$. For any real number $x$ we denote by $[x]$ the integral part of $x$, and by $\{x\}$ the fractional part of $x$, thus $0 \le x - [x] = \{x\} < 1$.

It is easily verified that

$$F(rz) = r^2 F(z),\ F(z_1 z_2) = -F(z_1)F(z_2),$$

$$F(F^k) = (-1)^{k-1} F^k(z) \text{ for } z, z_1, z_2 \text{ in } Q(\sqrt{D}),\ r \text{ in } q,\ k \text{ in } /N$$

PELL'S EQUATION is the diophantine equation

$$\boxed{|F(z)| = 1} \qquad \left(z \in /N_0\left(\sqrt{D}\right)\right)$$

"An especially notorious diophantine equation, in fact the issue of a celebrated challenge from Fermat to the English mathematicians of his time, is the equation $x^2 - dy^2 = 1$, when $d$ is a positive integer other than a perfect square. It is usually referred to as the Pell equation but the nomenclature, due to Euler, has no historical justification since Pell apparently made no contribution to the topic..."

See p. 74 in Alan Baker: A concise introduction to the theory of numbers Cambridge University Press (1984).

See also Barry Mazur: Bulletin Amer Math. Svc **14**, No. 2, April 1986 (pp. 208-209) and A. Weil: Number theory, an approach through

History, from Hammurapi to Legendre Birkhauser (1984).

The Pell equation was studied by Brahmagupta (7 Century A.D), Bhaskara (12 Century A.D). Narayana Pandita (1356 A.D) and others. But it is Bhaskara's Chakravala Method that gives an effective Algorithm for finding all the solutions of the Pell equation.

The purpose of this supplement is to present a self-contained treatment of the Bhaskara's Chakravala Method for students and teachers in our schools and colleges. I hope that this beautiful work of Bhaskara attracts their attention. The earlier versions given by me in ([1], [2]) were not satisfactory. The description of the Chakravala method which is being now presented in this supplementary is a greatly simplified and improved version which, I believe reflects more faithfully Bhaskava's own perception of the problem.

## § 2. THE BHASKARA COEFFICIENTS $m_\lambda$ AND $h_\lambda$ FOR $\sqrt{D}$

All our considerations stem from the following Lemma which plays a crucial role in our proofs.

**Lemma 1**

There exist in /N sequences $\{m_\lambda\}$ and $\{h_\lambda\}$ ($\lambda \geqq -2$) such that, for $\lambda \geqq 0$,

(1) $\boxed{D - m_{\lambda 2}^2 = h_{\lambda-2}h_{\lambda-1} \text{ with } 0 < \sqrt{D} - m_{\lambda-2} < h_{\lambda-2} < \sqrt{D} + m_{\lambda-2}}$

**Proof:** (by induction on $\lambda$) let

$$d_0 = \left[\sqrt{D}\right],\ m_{-2} = d_0,\ h_{-2} = D - d_0^2,\ h_{-1} = 1.$$

Since $0 < \sqrt{D} - d_0 < 1 < \sqrt{D} + d_0$ thus

$0 < \sqrt{D} - d_0 < D - d_0^2 < \sqrt{D} + d_0 \left(1_\lambda\right)$ is valid for $\lambda = 0$. Assure ($L\ \lambda$) and put.

(2) $\boxed{\delta_\lambda = \dfrac{\sqrt{D + m_{\lambda-2}}}{h_{\lambda-1}} \text{ and } \Delta_\lambda = \dfrac{\sqrt{D + m_{\lambda-1}}}{h_{\lambda-1}}}$, $(\lambda > 0)$

we *define* $m_{\lambda-1}$ as follows:

(3) $\boxed{m_{\lambda-1} \equiv -m_{\lambda-2}\left(\operatorname{mod} h_{\lambda-1}\right)}$ such that $\boxed{m_{\lambda-1} < \sqrt{D} < m_{\lambda-1} + h_{\lambda-1}}$

$m_{\lambda-1}$ is thus the well-defined term in the arithmetic progression with initial term $-m_{\lambda-2}$ and common difference $h_{\lambda-1}$ satisfying the condition $m_{\lambda-1} < \sqrt{D} < m_{\lambda-1} + h_{\lambda-1}$. Put.

$$(4) \quad a_\lambda = \frac{m_{\lambda-1} + m_{\lambda-2}}{h_{\lambda-1}} \qquad (x \geqq 0)$$

Thus $a_\lambda$ is an *integer* we have, from (3), $0 < \dfrac{\sqrt{D} - m_{\lambda-1}}{h_{\lambda-1}} < 1$,

while, from (2) and (4), $\delta_\lambda - a_\lambda = \dfrac{\sqrt{D} - m_{\lambda-1}}{h_{\lambda-1}}$.

Hence $0 < \delta_{\lambda-a\lambda} < 1$, in other words

$G_\lambda = [\delta_\lambda]$. But, by $(1_\lambda)$, $\delta_\lambda = \dfrac{h_{\lambda-2}}{\sqrt{D} - m_{\lambda-2}} > 1$, therefore $G_\lambda \underline{Z}\ 1$.

It follows from (4) that $m_{\lambda-1} + m_{\lambda-2} \underline{Z}\ h_{\lambda-1} > \sqrt{D} - m_{\lambda-1}$

so that $2m_{\lambda-1} > \sqrt{D} - m_{\lambda-2} > 0$ (by inductive assumption $1_\lambda$)

$m_{\lambda-1}$ is therefore a *positive* integer, $m_{\lambda-1}$ is in /N.

We have, from (3), $m_{\lambda-1} \equiv -m_{\lambda-2}$ (mod $h_{>-1}$), $m_{\lambda-1}^2 \equiv m_{2\lambda-2}$ $(\text{mod}\, h_{\lambda-1})$, $D - m_{\lambda-1}^2 \equiv \pm - m_{\lambda-2}^2 \equiv h_{\lambda-2}h_{\lambda-1} \equiv \ 0(\text{mod}\, h_{\lambda-1})$,

(prom $(1_\lambda)$), thus $\dfrac{D - m_{\lambda-1}^2}{h_{\lambda-1}}$ is an integer, which will be positive

since $0 < m_{\lambda-1} < \sqrt{D}$. We define

$$h_\lambda = \frac{D - m_{\lambda-1}^2}{h_{\lambda-1}};\ \text{thus } h_\lambda \text{ is in } /N.$$

Next, from $(1_\lambda)$, (2) and (4), $\Delta\lambda - G_\lambda = \dfrac{\sqrt{D} - m_{\lambda-1}}{h_{\lambda-1}} > 0$, thus

$\Delta_\lambda > G_\lambda \geqq 1$, $\Delta_\lambda > 1$, i.e. $h_{\lambda-1} < \sqrt{D} + m_{\lambda-1}$ consequently, you have found $m_{\lambda-1}$ and $h\lambda$ in /N satisfying the conditions.

$D - m_{\lambda-1}^2 = h_{\lambda-1} h_\lambda$ with $0 < \sqrt{D} - m_{\lambda-1} < h_{\lambda-1} < \sqrt{D} + m_{\lambda-1}$ which is precisely $(1_{\lambda+1})$. The induction is complete and the Lemma is established.

Since $\Delta_\lambda = a_\lambda + \dfrac{\sqrt{D} - m_{\lambda-2}}{h_{\lambda-1}} = a_\lambda + \dfrac{h_{\lambda-2}}{\sqrt{D} +_{m-2}}$

$$= a_\lambda + \frac{1}{\Delta_{\lambda-1}}, \text{ where } \Delta_{\lambda-1} > 1, \left(\text{See } (1_\lambda)\right),$$

We must have $[\Delta_\lambda] = a_\lambda$ and $\{\Delta_\lambda\} = \dfrac{1}{\Delta_{\lambda-1}}$

Again $\delta_\lambda = a_\lambda + \dfrac{\sqrt{D} - m_{\lambda-2}}{h_{\lambda-1}} = a_\lambda + \dfrac{h_\lambda}{\sqrt{D} +_{m-1}}$

$$= a_\lambda + \frac{1}{\delta_{\lambda+1}}, \text{ therefore } \{\delta_\lambda\} = \frac{1}{\delta_{\lambda+1}}, \text{ consequently}$$

(5) $\boxed{[\Delta_\lambda] = [\delta_\lambda] = a_\lambda}$ and $\boxed{\{\Delta_\lambda\} = \dfrac{1}{\Delta_{\lambda-1}}, \{\delta_\lambda\} = \dfrac{1}{\delta_{\lambda+1}}}$

We have $\bar{\delta}_{\lambda+1}\Delta_\lambda = \dfrac{-\sqrt{D} + m_{\lambda-1}}{h_\lambda}, \dfrac{\sqrt{D} + m_{\lambda-1}}{h_{\lambda-1}} = \dfrac{D - m2_{\lambda-1}}{h_\lambda\, h_{\lambda-1}} = -1,$

(6) $\boxed{\bar{\Delta}_\lambda = -\dfrac{1}{\bar{\delta}_{\lambda+1}}}$

We see that

(7) $\Delta_\lambda = \Delta_\lambda \Leftrightarrow \delta_{\lambda+1} = \delta_{\lambda+1}$

We have from (5) $\Delta_{\lambda+1} = \Delta_{\lambda+1} \Rightarrow \{\Delta_{\lambda+1}\} = \{\Delta'_{\lambda+1}\} \Rightarrow \Delta_\lambda = \Delta_\lambda'$. Conversely, if $\Delta_\lambda = \Delta_\lambda'$, then from (7), $\delta_{\lambda+1} = \delta_{\lambda'+1}$, $\{\delta_{\lambda+1}\} = \{\delta_{\lambda'+1}\}$, and therefore from (5), $\delta_{\lambda+2} = \delta_{\lambda'+2}$ which implies $\Delta_{\lambda+1} = \Delta_{\lambda'+1}$ (See 17) consequently.

(8) $\boxed{\Delta_\lambda = \Delta_\lambda, \longleftrightarrow \Delta_{\lambda+1} = \Delta_{\lambda'+1}}$

Since we always have $0 < \sqrt{D} - m_\lambda < \sqrt{D} + m_\lambda < h_\lambda$ $< \sqrt{D} + m_\lambda$ $(\lambda \geqq -2)$ we obtain

(9) $\boxed{0 < m_\lambda < \sqrt{D} \text{ and } 0 < h_\lambda < 2\sqrt{D}}$

Since the number of positive integers which are $< \sqrt{D}$, (resp. $< 2\sqrt{D}$, is necessarily finite it follows that in the sequences.

$$\{m_\lambda, \lambda_{\geqq 2}\}, \{h_\lambda, \lambda_{\geqq 2}\}, \{\Delta_\lambda, \lambda_{\geqq 0}\}$$

There can only exist finitely many distinct terms. There must exist α, β with $\beta \geqq \alpha + 1$ such that $\Delta_\alpha$ $\Delta_\beta$. But, then, by (5), we get $\Delta_{\alpha-1} = \Delta_{\beta-1}, ..., \Delta_0 = \Delta_{\beta-\alpha}$ $(\beta-\alpha \geqq 1)$. In other words, $\Delta_0$ must occur again in the sequence $\{\Delta_\lambda, \lambda \geqq 0\}$. Denote by $N$ the least positive integer such that $\Delta_0 = \Delta_N$, i.e.

(10) $\boxed{\Delta_0 = \Delta_N, \text{ and } \Delta_0 \neq \Delta_\lambda \text{ for } 0 <_\lambda < N}$

Owing to (8), $\Delta_0 = \Delta_N \Rightarrow \Delta_1 = \Delta_{N+1} \Rightarrow ... \Rightarrow \Delta_\lambda = \Delta_{N+\lambda}$ $(\lambda \geq 0)$. Changing $\lambda$ to $\lambda + N$ we get $\Delta_{N+\lambda} = \lambda_{ZN+\lambda}$ – and so on thus.

(11) $\Delta_\lambda = \Delta_{n+kN}$ $(\lambda \geqq 0, k \geqq 0)$

We prove: for $\lambda > 0$,

(12) $\boxed{\text{If } \Delta_\lambda = \Delta_0, \text{ then } \lambda = kN \text{ for some } k \geqq 1}$

Owing to (10), $\lambda \geqq$ N. Divide $\lambda$ by $N$ to write $\lambda = kN + \lambda 0$ with $0 \geqq \lambda_0 < N$. Then by (11) and (12).

$\Delta_0 = \Delta_\lambda = \Delta_{kN+\lambda 0}$, i.e. $\Delta_0 = \Delta_{\lambda 0}$. Here, because of (10), $\lambda 0$ must be = 0. Hence $\lambda = kN$ which proves (12).

## § 3. THE SEQUENCE $\{Z_\lambda, \lambda \geq 0\}$

We define the sequence $\{Z_\lambda, \lambda \geqq -1\}$ recursively by

(13) $\boxed{z_\lambda = \Delta_\lambda z_{\lambda-1}}$ $\boxed{(z_{-1} = 1)}$

(Note that $z_{-1} = \sqrt{D} + d_0$ since $F(\Delta_\lambda) = -\Delta_\lambda \bar{\Delta}_\lambda \Delta_\lambda$

$$= \frac{D - m_{\lambda-1}^2}{h_{\lambda-1}^2} = -\frac{h_\lambda}{h_{\lambda-1}} \text{ we have } F(z_\lambda) = -\frac{h_\lambda}{h_{\lambda-1}} F(z_{\lambda-1})$$

Whence follows

$$\frac{F(z_\lambda)}{h_\lambda} = -\frac{F(z_{\lambda-1})}{h_{\lambda-1}} = ... = (-1)^\lambda \frac{F(z_0)}{h_0} = (-1)^\lambda$$

Since $F(z_0) = D - d_0^2$ and $h_0 = \dfrac{D - m_{-1}^2}{h_{-1}} = D - d_0^2$, thus

(14) $\boxed{F(z_\lambda) = (-1)^\lambda h_\lambda}$

Now, from (5), we have

$$z_\lambda = \Delta_\lambda z_{\lambda-1} = \left(a_\lambda + \frac{1}{\Delta_{\lambda-1}}\right) z_{\lambda-1} = a_\lambda z_{\lambda-1} + z_{\lambda-2},$$

accordingly

(15) $\boxed{x_\lambda = a_\lambda x_{\lambda-1} + x_{\lambda-2} \text{ and } y_\lambda = a_\lambda y_{\lambda-1} + y_{\lambda-2}}$

Since $z_{-1} = 1$ and $z_0 = \sqrt{D} + d_0$, we have

$x_{-1} = 0, y_{-1} = 1$ and $x_0 = 1, y_0 - d_0$

Since $a_\lambda$ are all in /N, the relations (15) and (15) show that $x_\lambda$ and $y_\lambda$ are all in /N.

Put $\varepsilon_\lambda = x_\lambda{-1}\ y_\lambda - x_\lambda\ y_{\lambda-1}$, then

$\Sigma_\lambda = x_{\lambda-1}\ (a_\lambda y_{\lambda-1} + y_{\lambda-2}) - (a_\lambda x_{\lambda-1} + x_{\lambda-2})\ y_{\lambda-1}$

$= -x_{\lambda-2}\ y_{\lambda-1} + x_{\lambda-1}\ y_{\lambda-2} = -\Sigma_{\lambda-1},$

Whence follows $\varepsilon_\lambda = -\varepsilon_{\lambda-1} = ... = (-D^\lambda\ \Sigma_0 = -1)^{\lambda+1}$, i.e.

(16) $\boxed{x_{\lambda-1}\ y_\lambda - x_\lambda\ y_{\lambda-1} = (-1)^{\lambda+1}}$

Which implies that $(x_\lambda, y_\lambda) = 1\ (\lambda_{\geqq 0})$, thus

(17) $\boxed{z_\lambda \varepsilon / N_0(\sqrt{D})}$ $(\lambda\ \underline{Z}_0)$

We have from $\left(10, \Delta_0 = \Delta_N, \text{i.e. } \sqrt{D} + d_0 = \dfrac{\sqrt{D} + m_{N-1}}{h_{N-1}}\right)$

Therefore $h_{N-1} = 1$, and so from (14)

(18) We have found a solution $Z_{N-1}$ of the Pell equation

Since $F\left(Z_{N-1}^{k}\right) = (-1)^{k-1} F^{k}\left(Z_{N-1}\right)\left(K_{-1,2,\ldots}\right)$

We have

(19) $\boxed{\left|F\left(Z_{N-1}^{k}\right)\right| = 1}$ $\quad (K_{-1,2\cdots})$

i.e. $Z_{N-1}^{k}\left(k \geqq 1\right)$ are all solutions of the Pell equation.

We wish to show that there are no other solutions. For this purpose, we prove

**Lemma 2**

If $Z = \sqrt{D}\,x + y,\ x,\ y$ in $/N$, $\left(x,\ y\right) = 1$ and $z$ does not belong to the sequence $\left\{z_{\lambda},\ \lambda \geqq 0\right\}$ then

$$|F(z)| > 1$$

**Proof:** Since $z_{\lambda} = \Delta_{\lambda} z_{\lambda-1} > z_{\lambda-1}\left(\lambda_{\geqq 0}\right)$,

$\left\{z_{\lambda},\ \lambda_{\geqq 0}\right\}$ is a *strictly* increasing sequence of positive numbers. Hence if $z > 0$ does not lie in the sequence there must exist a uniquely determined index $\lambda$ such that

(20) $\boxed{z_{\lambda-1} < z < z_{\lambda}}$

Solve for *A, B* the equations

$x = Ax_{\lambda-1} + B_{x\lambda},\ y = Ay_{\lambda-1} + By_{\lambda}$

Since the co-efficients are all in $/N$ and $\begin{vmatrix} x_{\lambda-1} & x_{\lambda} \\ y_{\lambda-1} & y_{\lambda} \end{vmatrix} = (-1)^{\lambda+1}$ if

follows that A and B must be *integers*.

Hence we have

$$z = A\,z_{\lambda} + B_{\lambda}\,z_{\lambda}\ (A,\ B \text{ integers})$$

We claim that

(21) $A \neq 0,\ B \neq 0$ and $\dfrac{B}{A} < 0$

This follows from (20). In fact,

$A = 0 \Rightarrow x = Bx_\lambda, y = By_\lambda \Rightarrow B = 1$ since $(x, y) = 1 = (x_\lambda, y_\lambda)$ which implies that $\Rightarrow z = z_{\lambda-1}$. Similarly $B = 0 \Rightarrow z = z_\lambda$. Again $A < 0$ and $B < 0 \Rightarrow z < 0$, while $A > 0$ and $B > 0 \Rightarrow z > Bz_\lambda \geqq z_\lambda$. The condition (20) precludes all these possibilities. Now

$$(22) \quad \bar{z} = A \cdot \bar{z}\lambda_{-1} + B\bar{z}_\lambda = A\bar{z}_{\lambda-1}\left(1 + \frac{B}{A}\bar{\Delta}_\lambda\right)$$

But $\bar{\Delta}\lambda = \dfrac{-\sqrt{D} + m_{\lambda-1}}{h_{\lambda-1}} < 0$ and $\dfrac{B}{A} < 0$, therefore $\dfrac{B}{A}\bar{\Delta}_\lambda > 0$.

Taking absolute values in (22) we obtain, accordingly,

$$(23) \quad |\bar{z}| > |A||\bar{z}_{\lambda-1}| \geqq |\bar{z}_{\lambda-1}|, \text{ thus } |\bar{z}| > |\bar{z}_{\lambda-1}|$$

since $|A|$, being a positive integer, is $\geqq 1$.

Consequently, from (20) and (23)

$$|F(Z)| = Z|\bar{z}| > z_{\lambda-1}|\bar{z}_{\lambda-1}| = F|(z_{\lambda-1})| = h_{\lambda-1} \geqq 1$$

i.e. $\quad |F(z)| > 1$.

The Lemma is proved.

The Lemma 2 means that no solutions of the Pell Equation can exist outside the sequence $\{z_\lambda, \lambda \geqq 0\}$.

**Remark.** Infact, a sharper result follows from (23):

$$z \notin \{z_\lambda, \lambda \geqq 0\} \Rightarrow |F(z)| > \sqrt{D}.$$

[See Remark 1°, P. 13].

We prove my induction in $\lambda$.

$$(24_\lambda) \quad \boxed{z_{N+\lambda} = z_\lambda z_{N-1}} \; \lambda \geqq -1$$

Since $z_{-1} = 1$, this is true for $\lambda = 1$. We show $(2\Delta_\lambda) \Rightarrow (2\Delta_{\lambda+1})$. We have $z_{N+\lambda+1} = \Delta_{N+\lambda+1}\ z_{N+\lambda} = \Delta_{\lambda+1}\ z_\lambda\ z_{N-1} = z_{\lambda+1}\ z_{N-1}$ whence follows $(2\Delta_{\lambda+1})$, (See (10) and (13)). As a consequence of $(2\Delta_\lambda)$ we prove that

$$(25_k) \quad \boxed{z_{kN+\lambda} = z_\lambda z_{N-1}^k} \qquad (k \geqq 0, \lambda \geqq -1)$$

In fact, put $\lambda' = kN + \lambda$, so that $(k + 1)\lambda = N + \lambda'$, and $z_{(k+1)}$ $N + \lambda = z_{N+\lambda'} = z_{\lambda'}, z_{N-1}$, by $(2\Delta_{\lambda'})$. By inductive assumption $(25_k)$.

$z_{\lambda'=z_{kN+\lambda}} = z_\lambda z_{N-1}^{k}$ consequently $z_{(k+1)N+\lambda} = z\lambda z_{N-1}^{k+1}$ where follows $(25_{k+1})$. Since $(25_k)$ is valid for $k = 0$, the induction is complete and $(25_k)$ is proved.

In particular, we have

(26) $$\boxed{z_{kN-1} = z_{N-1}^{k}} \qquad k \geqq 0$$

Suppose now that $z \in N_0\left(\sqrt{D}\right)$ satisfies the Pell equation. By Lemma 2, there exists $\lambda \geqq 0$ such that $Z = Z_\lambda$. Then $|F(Z\lambda| = |F(Z)| = 1$, i.e. $h_\lambda = 1$. Since $[\Delta_{\lambda+1}] = [\delta_{\lambda+1}] = a_{\lambda+1}$ (See (5)), we have

$$\left[\frac{\sqrt{D}+m_\lambda}{h\lambda}\right] = \left[\frac{\sqrt{D}+m_{\lambda-1}}{h_\lambda}\right] = \frac{m_\lambda + m_{\lambda-1}}{h_\lambda}$$

Therefore $[d_0 + m_\lambda] = [d_0 + m_{\lambda-1}] = m_\lambda + m_{\lambda-1}$

$$m_\lambda = m_{\lambda-1} = d_0$$

Consequently $\Delta_{\lambda+1} = \sqrt{D} + d_0 = \Delta_0$. But, then, by (12), $\lambda+1 = kN$ for some $k \geqq 1$, and by (26), $z_\lambda = z_{kN-1} = z_{N-1}^{k}$. We have proved.

**Theorem**

All the solutions of the Pell equation
$|F(z)| = 1, \qquad z \in z_0\left[\sqrt{D}\right]$
are given by $z = z_{N-1}^{k} \quad (k \geqq 1)$

**Remark 1°**

We can prove a slightly sharper result than Lemma 2. Solving for $B$ the equations

$$x = A\,x_{\lambda-1} + Bx_\lambda,\ y = Ay_{\lambda-1} + By_\lambda$$

We have

$$B(-1)^{\lambda+1} = \begin{vmatrix} x_{\lambda-1} & x \\ y_{\lambda-1} & y \end{vmatrix},\ B(-1)^{\lambda+1}\,2\sqrt{D} = \begin{vmatrix} \sqrt{D} & 1 \\ -\sqrt{D} & 1 \end{vmatrix} \begin{vmatrix} x_{\lambda-1} & x \\ y_{\lambda-1} & y \end{vmatrix}$$

$$= \begin{vmatrix} z_{\lambda-1} & z \\ \bar{z}_{\lambda-1} & \bar{z} \end{vmatrix}$$

Whence follows, owing to (23),

$$|B| \cdot 2\sqrt{D} \leqq z_{\lambda-1}|\bar{z}| + z|\bar{z}_{\lambda-1}| < z|\bar{z}| < z|\bar{z}| = 2|F(z)|$$

Since $|B| \geqq 1$, we conclude that

$$\boxed{z \notin \{z_\lambda, \lambda \geqq 0\} \Rightarrow |F(z)| > \sqrt{D}}$$

or what is the same, in order that $z \in \{z_\lambda, \lambda \geqq 0\}\left(z \in N_0\left[\sqrt{D}\right]\right)$ it is *sufficient* that $|F(z)| < \sqrt{D}$. On the other hand, if $z \in \{z_\lambda, \lambda \geqq 0\}$, then $z = z_\lambda$ for some $\lambda \geqq 0$, and $|F(z)| = |F(z_\lambda)| = h_\lambda < 2\sqrt{D}$ i.e. $|F(z)| < 2\sqrt{D}$ is a *necessary* condition.

However, $|F(z)| < 2\sqrt{D}$ is not a necessary condition.

In fact, using the recurrence relations (3) we finds by an easy computation.

$$D = 109,\ d_0 = \left[\sqrt{D}\right] = 10,\ |F(z_2)| = b_2 = 12 > \sqrt{D}$$

The problem is to find a necessary and sufficient condition.

**Question:** Is the condition $|F(z)| < 2\sqrt{D}$ sufficient?

**Remark 2°**

We show that

$$(13_\lambda) \qquad \boxed{\delta_\lambda = \Delta_{N-\lambda}} \qquad (0 \leq \lambda \leq N)$$

Since $\Delta_0 = \Delta_N$, (See (10)), and $\delta_0 = \sqrt{D} + d_0 = \Delta_0$ $(13_\lambda)$ holds for $\lambda = 0$. We prove (13λ) by induction on λ. Assuming $(13_\lambda)$ and taking fractional parts (See (5)), we get $\delta_{\lambda+1} = \Delta_{N-\lambda-1}$, i.e. $(13_{\lambda+1})$. Taking integral parts in $(13_\lambda)$ we get

$$a_\lambda = a_{N-\lambda} \qquad (0 \leqq \lambda \leqq N)$$

## SUPPLEMENT–III



### § 1. POLYNOMIAL RINGS

Our objective in this supplement III is to prove certain theorems which form a natural continuation from the results that we have already studied in this book.

We need to review briefly some basic results from Algebra. For further information the student is referred to:

1° **E. Hecke:** Lectures on the theory of algebraic numbers (Chapter IV) Springer Verlag (1981).

2° **I. Niven and H. Zuckerman:** Introduction to the theory of numbers (Chapter 9, pages 188-203).

Let **Z** denote the ring of rational integers {..., –2, –1, 0, 1, 2, ...} and **Q** the field of rational numbers $\left\{\frac{a}{b}, b \neq 0\right\}$, $\{a, b\} \subset \mathbf{Z}$ $\mathbf{Q}(x)$ denotes the ring of polynomials in $x$ with coefficients in **Q**, and by $\mathbf{Z}(x)$ the ring of polynomials with coefficients in **Z**. A polynomial $f(x)$, in $\mathbf{Q}(x)$, (respectively in $\mathbf{Z}(x)$ is said be *irreducible* if it cannot be expressed a product $f(x) = f_1(x) f_2(x)$ where $f_1(x), f_2(x)$ are in $\mathbf{Q}(x)$ (respectively in $\mathbf{Z}(x)$).

Given polynomial $f(x) = a_0x^n + a_1x^{n-1} + ... + a_n$ where $a_0 \neq 0$, we say that $n$ is the *degree* of $f(x)$ and $a_0$ the highest co-efficient of $f(x)$. If all the co-efficients $a_0, ..., a_n$ are zero, $f(x)$ is called a zero-polynomial, it has no degree. If deg $f(x) = 1$, then $f(x) = a_0$ is called a constant polynomial. We observe

(1) $\deg f_1(x) f_2(x) = \deg f_1(x) + \deg f_2(x)$

A polynomial $f(x)$ is said to be *monic* of its highest co-efficient is = 1.

**Gauss's Lemma:**

1° Let $f(x)$, $f_1(x)$, $f_2(x)$ be *monic* polynomials such that $f(x) = f_1(x) f_2(x)$, where $f(x) \in \mathbf{Z}(x)$, $f_i(x) \in \Phi(x)$ $(i\ 1, 2)$

then $f_1\ (x)$ and $f_2(x)$ have coefficients in **Z**.

2° If a monic polynomial $f(x) \in \mathbf{Z}(x)$ is the product of this monic polynomial in $\Phi(x)$, then, actually, $f_1(x)$ and $f_2(x) \in \mathbf{Z}(x)$.

It is not easy to settle whether a given polynomial $f(x) \in \mathbf{Z}(x)$ is irreducible or not. The following criterion is often useful.

**Theorem 1. (Einsenstein Irreducibility Criterion)**

Let $G(x) = g_0 + g_1x + ... + g_nx^n$ be in $\mathbf{Z}(x)$. Suppose there exists a *prime* number such that the following conditions are fulfilled:

(2) $g\ (0) \not\equiv 0 \pmod p$, $g_i \equiv 0 \pmod p$, $(i = 0, 1, ..., p-1)$ and $g_0 \not\equiv 0 \pmod{p^2}$. Then $G(x)$ is irreducible.

**Proof:** Suppose on the contrary, that

$$G(x) = A(x)B(x) \text{ where } \left| \begin{array}{l} A(x) = a_0 + a_1x + ... + g_rx^r \\ B(x) = b_0 + b_1x + ... + b_sx^s \end{array} \right.$$

where $r > 0$, $s > 0$, $n = r + s$.

We have $g_0 = a_0b_0$. Since, by (2), $g_0 \not\equiv 0 \pmod p$, we have $a_0b_0 \equiv 0 \pmod p$. Since $p$ is prime, either $a_0 \equiv 0 \pmod p$ or $b_0 \equiv 0 \pmod p$. However both cannot hold, since then $a_0b_0 \equiv 0 \pmod{p^2}$, i.e. $g_0 \equiv 0 \pmod{p^2}$ contradicting the hypothesis (2). Assume then

(3) $b_0 \not\equiv 0 \pmod p$.

All the coefficients of $A(x)$ cannot be divisible by $p$, since then $A(x)$ can be written as $A(a) = p\,A'(x)$ so that $a(x) = pA'(x)\ B(x)$. Hence $p$ will have to divide all the co-efficients of $a(x)$ violating the hypothesis that $g_0 \not\equiv 0 \pmod p$. We choose in the sequence $a_0$, $a_1$ ... $a_n$ the very first that is $\not\equiv 0 \pmod p$, i.e.

(4) $a_0, a_1, ...\ a_{i-1}\ a_{re} \equiv 0 \pmod p$ but $a_i \not\equiv 0 \pmod p$. We have $g_i = a_ib_0 + a_{i-1}\ b_1 + ... + a_ib_0$. By hypothesis $g_i \equiv 0$ (mod p) for $i < n$. In this manner we obtain, in virtue of (4)

$$a_i b_0 \equiv 0 \pmod p.$$

But this is impossible because, by (3) and (4) $a_i \not\equiv 0 \pmod p$ and $b_0 \not\equiv 0 \pmod p$.

This contradiction establishes the criterion.

**Example 1.** For any prime $p$, the polynomial $x^n - p$ is irreducible. The conditions of the criterion are obviously fulfilled.

**Example 2.** The polynomial

$$C(x) = x^{p-1} + x^{p-2} + ... + 1$$

is irreducible.

Consider $f(x) = C(x + 1)$. Since $C(x) = \dfrac{x^p - 1}{x - 1}$,

we have $f(x) = \dfrac{1}{x}\left[(x+1)^p - 1\right]$ *i.e.*,

$$f(x) = \binom{p}{0} + \binom{p}{1}x + ... + \binom{p}{p-1}x^{p-2} + x^{p-1}$$

For $1 \le k \le p$-1, the binomial co-efficient

$$\binom{p}{k} \text{ no } \binom{p}{k} = \frac{p(p-1)...(p-k+1)}{k!}.$$

Hence $k!\binom{p}{k} \equiv 0 \pmod p$, but $k! \not\equiv 0 \pmod p$ because $1 \le k \le p-1$.

Consequently $\binom{p}{k} \equiv 0 \pmod p$, $(0 < k < p)$. The conditions of the Eisenstein criterion are satisfied. $f(x)$ is irreducible. Consequently $C(x)$ is irreducible, otherwise a factorization yields one for $f(x)$ also.

## § 2. STATEMENT OF THE FUNDAMENTAL THEOREM ON SYMMETRIC FUNCTIONS

In order to prove that the algebraic numbers form a field we need this theorem. Let $x_1, x_2, ... x_n$, be independent variables we consider the polynomial in $t$:

(1) $f(t) = (t - x_1)(t - x_2)...(t - x_n)$

The coefficients $\sigma_1, \dots \sigma_n$ appearing are called *elementary symmetric functions* of $x_1, \dots x_n$. We have $\sigma_1 = \sum_{i=1}^{n} x_i$, $\sigma_2 = \sum x_i x_j$, $(1 \le i \le n) \dots \sigma_n = x_1 x_2 \dots x_n$, thus

(2) $\boxed{\sigma k = \sum x_{i_1} x_{i_2} \dots x_{i_k}, (1 \le i_1 < \dots < i_k \le m)}$

$\sigma_k$ remain invariant under every permutation of (1, 2, ..., $n$), i.e.

(3) $\sigma_k(x_{i1}, \dots, x_{in}) = \sigma_k(x_1, \dots, x_n)$

**Theorem A**

Given a *symmetric* polynomial $S(x_1, \dots x_n)$ we can find a polynomial $G(t_1, \dots t_n)$ such that

(4) $\boxed{S(x_1, \dots, x_n) = G(\sigma_1, \dots, \sigma_n)}$

The coefficients of $G$ can be computed out of the co-efficients of $S$ through the operations of addition, subtraction and multiplication.

**Theorem B**

Suppose $S(x_1, \dots, x_n \mid y_1, \dots y_m)$ is a polynomial which is symmetric in $(x_1, \dots x_n$ and separately in $(y_1, \dots, y_m)$, i.e. $S(x_{i_l}, \dots, x_{i_n} \mid y_{i_1}, \dots, y_{i_m}) = S(x_1, \dots, x_n \mid y_1, \dots, y_m)$ for all permutations $(i_l, \dots, i_n)$ of (1, 2, ..., $n$) and all permutations $(i_1, \dots, i_m)$ of (1, 2, ..., $m$). Then there exists a polynomial

$G(m_1, \dots u_n \mid \vartheta_1, \dots, \upsilon_m)$ of $(n + m)$ variables such that

$S(x_1, \dots, x_m \mid y_1, \dots, y_m) = G(\sigma_1, \dots, \sigma_n \mid \tau_1, \dots, \tau_m)$

(See Hecke, Chapter IV, § 18).

## § 3. THE FIELD $Q$ OF ALGEBRAIC NUMBERS

**Definion 1.** A real or complex number $\vartheta$ is called an *algebraic number* over **Q** or simply an algebraic number if there exists a polynomial $f(x) \in \Phi(x)$ such that $f(\vartheta) = 0$.

(1) $f(z) = a_0 z^m + a_1 z^{n-1} + a_n z$, and $f(\vartheta) = 0$, $\{a_0, a_1, a_2, \dots a_n\} \subset \mathbf{Q}$

$$a_0 + a_1 \vartheta + \dots + a_n \vartheta_n = 0$$

$\vartheta$ is called a *root* (or *zero*) of $f(z)$. Among all the polynomials, which possess $\vartheta$ as a zero, let us choose a *monic* polynomial in

$Q(z)$ which has $\vartheta$ as a zero. We denote it by $f^*(z)$. Let $d$ be the degree of the polynomial $f^*(z)$. (Thus

(2) $f^*(\vartheta) = 0$, $d = \deg f^*(z)$ *minimal* $f^*(z) = z^d + a_n z^{d-1} + \ldots a_d$

By the very definition, $f^*(z)$ must be *irreducible,* since otherwise

$$f^*(z) = g_1(z)\, g_2(z),\ 0 < \deg g_i < d\ (i = 1, 2)$$

will yield $g_1(\vartheta) = 0$ or $g_2(\vartheta) = 0$, which violates the minimality of $d$.

Let, now, $F(z)$ be any polynomial in $Q(z)$ such that $F(\vartheta) = 0$, then $\deg F(z) > d$. Divide $F(z)$ by $1 f^*(z)$ to have

$F(z) = q(z) f^*(z) + r(z)$. Putting $z = \vartheta$ we find $r(\vartheta) = 0$, $0 \leq \deg r(z) < d$.

Here $\deg r(z)$ cannot be $> 0$, again because of the minimality of $d$. Thus $r(z)$ is the zero-polynomial. Consequently

(3) $f^*(z) | F(z)$, i.e. $f^*(z)$ divides $F(z)$

Suppose $F(\vartheta) \neq 0$, and let $d\ (z)$ be the g.c.d. of $F(z)$ and $f^*(z)$. Since $f^*(z)$ is irreducible, and minic, $d\ (z) \equiv 1$. Consequently g.c.d. of $F(z)$ and $f^*(z)$ must be 1, i.e. they are relatively prime: Accordingly we have:

$$(4)\quad \boxed{\begin{array}{l} F(\vartheta) = 0 \ \Rightarrow f^*(z) | F(z) \\ F(\vartheta) \neq 0 \ \Rightarrow \left(F(z)\, f^*(z)\right) = 1 \end{array}}$$

Consequently, we can find $A(z)$ and $B(z)$ in $Q(z)$ such that

$A(z)F(z) + B(z) f^*(z) \equiv 1$

But $z = \vartheta$ to find $A(\vartheta)\, F(\vartheta) = 1$, i.e.

$$(5)\quad \frac{1}{F(\vartheta)} = A(\vartheta) \qquad \deg A(z) \leq d$$

Now we wish to prove that the set of algebraic numbers over $\Phi$ forms a field i.e. if $\alpha$ and $\beta$ are in $\Phi$, then

$$\alpha + \beta,\ \alpha - \beta,\ \alpha \cdot \beta, \text{ and when } \beta \neq 0 \text{ also } \frac{\alpha}{\beta} \text{ are also in } \mathbf{Q}.$$

Let $\alpha$ be an algebraic number of degree $n$, and let $A(x)$ be its minimal polynomial. If $\beta$ is an algebraic number of degree $m$, $\beta_1, \ldots, \beta_n$ its conjugates, we have

$A(\alpha) = 0$, $A(\alpha) = (x - \alpha_1)(x - \alpha_2) \ldots (x - \alpha_n)$, $(\alpha = \alpha_3)$

Form the product

$S(x) = A(x - \beta_1)\, A(x - \beta_2) \dots A(x - \beta_m)$

Obviously the product $S(x)$ is symmetric in $(\alpha_1, \dots \alpha_n)$, and separately in $(\beta_1, \dots \beta_n)$. Herein

$$S(\alpha_i + \beta_i) = \prod_{j=1}^{m} A(\alpha_i + \beta_i - \beta_i) = 0,$$

since the product has its $j^{\text{th}}$ factor equal to $A(\alpha_i + \beta_i - \beta_i) = A(\alpha_i) = 0$. Hence $\alpha + \beta \in \bar{Q}$.

Next we show $\alpha\beta$ is algebraic.

We consider the product

$$P(x) = (\beta, \dots, \beta_m)^m \cdot A\left(\frac{x}{\beta_1}\right) \dots \left(\frac{x}{\beta_m}\right)$$

Again, obviously, $P(x)$ is symmetric in $\alpha_1, \dots, \alpha_m$, and separately in $\beta_1, \dots \beta_m$. Accordingly $P(x) \in Q(x)$. Also

$P(\alpha_i, \beta_j) = 0, \qquad (i = 1, \dots, n \text{ and } j = 1, \dots, n)$

This proves the assertion.

Let $\qquad \alpha \neq 0, \deg \alpha = n.$

$$f^*(\alpha) = 0.\, f^*(x) = x^n + a_1 x^{n-1} + \dots + a_n = 0$$

**Remark.** The constant term of a minimal polynomial can never be equal to zero.

In fact, if $f^*(x)$ is the minimal polynomial of $\vartheta$.

$$f^* = x^n + a_1 x^{n-1} + \dots + a_m = 0, \qquad (a_i \in Q)$$

$$(i = 1, 2, \dots, n)$$

If $a_m = 0$, $f^*$ ceases to be irreducible.

## § 4. TAYLOR EXPANSION FOR POLYNOMIALS

We begin by considering an example. Given the polynomial $a_0 + a_1 z + a_2 z^2$ and $\vartheta$ we want to determine $A_0, A_1, A_2$ such that the following identity holds.

(1) $a_0 + a_1 z + a_2 z^2 = A_0 + A_1(z - \vartheta) + A_2(z - \vartheta)^2$

Comparing like powers of $z$ in (1), we have

(3) $A_0 + A_1\vartheta + A_2\vartheta^2 = a_0$

$A_1 - 2A_2\vartheta = a_1$

$A_2 = a_2$

In order to determine the unknowns $A_0$, $A_1$, $A_2$ we have three homogeneous equations whose determinant has 1 along its diagonal, and 0 below the diagonal. Hence it is $\neq 0$, in fact $= 1$. Consequently the system has a *unique* solution for $A_0, A_1, A_2$. In general, we prove the following:

**Theorem 1 (Taylor expansion Theorem)**

Let $f(z)$ be a polynomial in $z$ of degree $n$ with real or complex coefficients

$\therefore$ $f(z) \in C(z)$ and $\vartheta \in C$. Then

$$f(z) = \sum_{k=0}^{n} \frac{f^{(k)}(\vartheta)}{k!} \cdot (z-\vartheta)^k \tag{4}$$

**Proof:** Let us write

$f(z) = A_0 + A_1 (z - \vartheta) + ... + A_n(z - \vartheta)^n$. Putting $z = \vartheta$ we find $f(\vartheta) = A_0$, differentiating w.r.t. $z$ and putting $z = \vartheta$ we find $f'(\vartheta) = A_1$, and so on. In general,

$$f^{(k)}(\vartheta) = k!A_k,\ i.e.,\ A_k = \frac{f^{(k)}(\vartheta)}{k!}$$

The theorem is proved.

**Theorem 2**

$$\text{Let } f(x) = \sum_{k=0}^{n} a_k \frac{x_k}{k!}, \left(\frac{a_k}{k!} \in Z\right)$$

(5) $F(x) = f(x) + f'(x) + ... + f^{(n)}(x)$

Assume $f^*(x)$ is the minimal polynomial of an algebraic number $\vartheta$. Let $H(\alpha)$ be the *height* of $\vartheta$, i.e.

$$H(\alpha) = \text{Max}\left\{\frac{|ak|}{k!}\right\}, (a_0 = 1),\ 0 \leq k \leq n \tag{6}$$

Then the following inequality holds:

For $|x - \vartheta| < 1$,

$$\begin{array}{ll} 1^\circ & |f(x)| \leqq |a-\vartheta| \cdot |F(x)| \\ 2^\circ & |F(x)| < \left| \dfrac{1}{\dfrac{1-|x|}{|x|^{n-1}}} \right| \quad \text{for } |x| < 1 \end{array} \tag{7}$$

**Proof:** We have

(8)

$$f(x) = a_0 + a_1 \frac{x}{\underline{1!}} + \ldots + a_n \frac{x^n}{\underline{n!}}$$

$$f'(x) = a_1 + \ldots + a_n \frac{x^{x-1}}{\underline{(n-1)!}} \qquad F(x) = a_0 + a_1\sigma_1(x) + \ldots + a_n\sigma_n(x)$$

$$\vdots \qquad \text{where}$$

$$f^{(n)}(x) = \qquad a_n \qquad \sigma_k(x) = 1 + x + x^k + \ldots + x^k$$

By Taylor's of $f(x)$ at $\vartheta$ we have:

$$f(x) = f(\vartheta) + (x - \vartheta) f'(\vartheta) + \ldots + \frac{(x-\vartheta)^n}{n!} f_{(\vartheta)}^{(n)}$$

Since $f(\vartheta) = 0$, this gives, for $(x - \vartheta) < 1$,

$$|f(x)| \leqq |x - \vartheta| \left[ f(\vartheta) + f'(\vartheta) + \ldots + f_{(\vartheta)}^{(n)} \right]$$

whence follows 1° in (9).

We estimate $|F(x)|$ under the condition $|x - \vartheta| < 1$. Since, by (8),

$$a_0 = 1, |a_k| \leqq k! H(\alpha) \qquad (k = 1, \ldots, n)$$

$$< n!\, H(\alpha)$$

We have

(9) $$|F(x)| \leqq H(\alpha) \cdot n! \left[ 1 + |\sigma_1(x)| + \ldots + |\sigma_n(x)| \right]$$

Put $|x| = \xi$, then $|\sigma_k(x)| \leqq \sigma_k(\xi) < \sigma_n(\xi)$.

Also we have that

$$\sigma_n(\xi) = 1 + \xi + \xi^2 + \ldots + \xi^n < \begin{cases} \dfrac{1}{1-\xi} & \text{for } 0 < \xi < 1 \\[2ex] \xi^n \dfrac{1}{1-\xi} & \text{for } \xi > 1 \end{cases}$$

Consequently

$$|F(a)| < H(\alpha) n! [1 + n\, \sigma_n(\xi)]$$

Now $|f(x) < |x - \vartheta|. L(\vartheta)$, where $L(\vartheta)$ is defined by

$$(10) \quad L(\vartheta) = \begin{vmatrix} H(\vartheta)n!\left(1+\dfrac{n}{1-\xi}\right) \text{ for } 0<\xi<1 \\ H(\vartheta)n!\left(1+\dfrac{n\xi^{n-1}}{\xi-1}\right) \text{ for } \xi>1 \end{vmatrix}$$

Herein $\xi = |x| + (x - \vartheta) < 1 + |\vartheta|$.

Finally we obtain the inequality

$$(11) \quad \boxed{1 > |x - \vartheta| > \frac{1}{L(\vartheta)}|f(x)|}$$

Now, $f(x)$ being the minimal polynomial of the algebraic number $\vartheta$, deg $\vartheta > 1$, we must have, for any rational number $p/q$, $f(p/q) \neq 0$. Here

$$f(x) = \sum_{k=0}^{n} a_k \frac{x^k}{k!} \in Z(x);\ i.e.,\ A_k = \frac{a_k}{k!} \in Z$$

Since $f(p/q) = A_0 + A_1\, p/q + ... + An\, p^n/q^n$, we have, consequently: $q^n f(p/q)$ is an *integer* $\neq 0$ thus its absolute value is $\geqq 1$.

$$1 > \left|\frac{p}{q} - \vartheta\right| > \frac{1}{L(\vartheta)} \cdot \frac{1}{q^d}$$

In this manner we have obtained

**Liouville's theorem**

If $\vartheta$ is an algebraic number of degree $p^n > 1$ there holds the inequality

$$12. \quad \boxed{\left|\vartheta - \frac{p}{q}\right| > \frac{1}{L(\vartheta)} \cdot \frac{1}{q^n}}$$

## § 5. TRANSCENDENCE OF *e* AND π

We have defined the field **Q** of algebraic numbers. We have the *ring* of *integers* **Z** in **Q**. We want to introduce the notion of *integral* elements in **Q**.

**Definition** (See Hecke: Chapter V, §§ 1-2)

An algebraic number $\vartheta$ of degree $n$ is said to be on *algebraic integer* if its irreducible, minimal, monic polynomial has also its coefficients as rational integers.

Thus, if $\vartheta$ is an algebraic integer,

$$f(\vartheta) = 0, \text{ where } f(x) = x^n + a_1x^{n-1} + ... + a_n,$$
$$a_k \in \mathbf{Z}(k = 1, ..., n)$$

If $\alpha$, $\beta$ are algebraic integers, then

$$\alpha + \beta, \alpha - \beta, \alpha \cdot \beta$$

are all algebraic integers.

If we write, $(\vartheta = \vartheta_1)$.

$$f(x) = (x - \vartheta_1) ... (x - \vartheta_n) = x^n - \sigma_1 x^{n-1} + ... + (-1)^n \sigma_n$$

we must have here $\sigma_1, ..., \sigma_n$ to be rational integers:

The *symmetric* functions of the *conjugates* of an algebraic *integer* are rational integers.

**Theorem 1**

If $\vartheta$ is an algebraic number, there exists a positive integer $\ell$ such that $\ell_\vartheta$ is an algebraic *integer*.

**Proof:** By hypothesis, $\vartheta$ satisfies a *monic* polynomial with rational co-efficients

$$x^n + \frac{a_1}{b_1} x^{n-1} + ... + \frac{a_n}{b_n} = 0$$

Let $\ell$ be the l.c.m. of the denominators $b_1, b_2, ..., b_n$. We have

$$\ell^n \left( x^n + \frac{a_1}{b_1} x^{n-1} + ... + \frac{a_n}{b_n} \right) = (b)^n + ab_0 \cdot a_1 x^{n-1} + ... + bb_m \ell^{n-1} a_1$$

which shows that $\ell_\vartheta$ satisfies a *monic* polynomial with rational integral coefficients. Consequently, $\ell_\vartheta$ is an algebraic integer. We also have that:

| The symmetric functions of the conjugates of an algebraic integer are all rational integers. |
|---|

We now proceed to the proof of transcendence of $e$ and $\pi$.

**Lemma 1°**

Let $f(x) \in \mathbf{Z}(x)$, $n = \deg f(x)$, and

(1) $f(x) = x^{p-1}(C_0 + C_1 x + ... + C_{n-p+1} x^{n-p+1})$.

where $C_k \xi \mathbf{Z}(k = 0, 1, ...n - p + 1)$. Let

$$F(x) = \sum_{k=0}^{n} f^{(k)}(x). \text{ Then}$$

(2) $$\boxed{\frac{1}{(p-1)|} F(0) \equiv C_0 \pmod{p}}$$

**Proof:** We find, by Taylor expansion at $x = 0$,

$$\sum_{k=0}^{n} \frac{f^{(k)}(0)}{k!} = x^{p-1}\left(C_0 + C_1 x + ... + C_{n-p+1}\, x^{n-p+1}\right)$$

Accordingly $f^{(k)}(0) = 0$, $(0 \leq b < p - 1)$.

(3) $$\frac{1}{(p-1)!} f^{(p-1)}(0) = C_0, \text{ and } \frac{f^{(p+v)}(0)}{(p+2)!} = C_{v+1}\,(v \geqq 0).$$

Therefore

$$\frac{1}{(p-1)!} f^{(p+v)}(0) = C_{v+1} \frac{(p+v)!}{(p-1)!} = C_{v+1} p(p+1)...(p+v)$$

(4) i.e. $$\frac{1}{(p-1)!} f^{(p+v)}(0) \equiv 0 \pmod{p}$$

**Lemma 2°**

Let $a \in \mathbf{Z}$, and $f(x) = (x - a)^p\,[C_0 + C_1(x - a) + ... + c_{n-p}x^{n-p}]$

Then

$$\boxed{\frac{1}{(p-1)!} F(a) \equiv 0 \pmod{p}}$$

**Proof:** We are by Taylor expansion at $x = a$

$$\sum \frac{f^k(Q)}{k!}(x-a)^k$$

$$= (x-a)^p \left[C_0 + C_1(x-a) + ... + C_{n-p}(x-0)^{n-p}\right]$$

Accordingly, $f^{C_n}(a) = 0, \left(0 \leq k \leq p\right)$ and

$$\frac{f^{(p+\nu)}(a)}{(p-1)!}C_{n-p}\frac{(p+\nu)!}{(p-1)!}=C_{n-p}p(p+1)\dots(p+\nu) \text{ which is } \equiv 0$$

(mod $p$). This proves the lemma.

Lemmas 1° and 2° yield us then

**Theorem 4**

Let $\alpha_1, \dots, \alpha_n$ be non-zero algebraic *integers* such that their elementary symmetric functions are *rational integers*. Let

(1) $f(x) = x^{p-1}(x - \alpha_1)^p \dots (x - \alpha_n)^p$, and $F(r) = \sum_{k-1}^{N} f^{(k)}(x)$,

($N = \deg f(x) = (n + 1)p - 1$). Then, with $C_0 = (-1)^p \alpha_1, \dots \alpha_n$,

(2) $\boxed{1° \frac{1}{(p-1)!}F(0) \equiv C_0 \pmod p}$

$$2° \frac{1}{(p-1)!}\left[F(\alpha_1)+\dots+F(\alpha_n)\right] - \equiv 0 \pmod p$$

**Proof:** 1° By Taylor expansion at $x = 0$: we have, putting

$g(x) = (x = \alpha_1)^p \dots (x - \alpha_n)^p$

$$\sum_{k=0} \frac{f^{(n)}(0)}{k!}x^k = x^{p-1}\sum_{\nu=0}^{N-p+1}\frac{g^{(\nu)}(0)}{\nu!}x^\nu$$

whence follows, (observing that $g(x) \in \mathbf{Z}(x)$) $f^{(k)}(0) = 0$, ($0 \le k < p-1$),

$$f^{(p-1)}(0) = \frac{(p+\nu)!}{\nu!}. \text{ Therefore } \frac{f^{(p+\nu)}(0)}{(p-1)!} = p \cdot \binom{p+\nu}{\nu} = \pmod p$$

and we have that $F(0) \equiv C_0 \pmod p$ which proves 1° in (6).

(3) 2° We now write

$f(x) = (x - \alpha_k)\, p \cdot g_k(x)$, ($k = 1, 2, \dots m$)

We use the Taylor expansion at $x = \alpha_k$:

$$\sum_{k=0}^{N}\frac{f^{(k)}(\alpha_k)}{k!} = (x-a_k)^p \cdot \sum_{\nu=0}^{N-p}\frac{g_k^{(\nu)}(a_k)}{\nu!}$$

which yields:

$$f^{(k)}(a_k)=0,\ (0\le k\le p),\ f^{(p+\nu)}(\alpha_k)=\frac{(p+\nu)!}{\nu!}\nu\geqq 0$$

i.e.

$$\frac{1}{(p-1)!}p\cdot\frac{(p+\nu)!}{p!\nu!}=p\cdot\binom{p+\nu}{\nu}\equiv 0\ (\text{mod}\ p)$$

Hence follows that

$$(4)\quad \boxed{\sum_{(k=1)!}^{n}\left[F(\alpha_1)+...+F(\alpha_k)\right]=\Sigma\binom{p+\nu}{\nu}g_k^{(\nu)}(\alpha_k)}$$

We observe that $g_0(x)$ was is $Z(x)$, but $g_k(x)$, ($i$ = 1, ..., $n$), are not. But the sum on the right hand side of (4) is a rational integral linear combination of the polynomials $g_k(x)$ which is symmetric in $\alpha_1$, ..., $\alpha_n$. By hypothesis, these symmetric functions are rational integers, and since the coefficients $\binom{p+\nu}{\nu}$ are all $\equiv 0$ (mod $p$), we obtain that, from (4) that

$$F(\alpha_1)+...+F(\alpha_n\equiv 0\ (\text{mod}\ p)$$

which proves 2° in (2), and establishes the theorem.

We shall use Theorem 3 and Theorem 4 to prove the transcendence of $e$ and $\pi$.

**Transcendence of *e* and π**

$$(5)\quad \sum_{k=0}^{m}g_k e^k=0$$

In order to prove the transcendence of $e$ we choose the auxiliary function

$$(1)\quad \boxed{f(x)=x^{p-1}(x-1)...(x-n)^p}$$

which we have considered above.

Suppose, on the contrary, that $e$ is algebraic, i.e. $e$ is a root of some polynomial $G(x)\in\mathbf{Z}(x)$.

(2) $\sum_{k=0}^{n} g_k e^k = 0, \qquad G(x) = \sum_{k=0}^{m} g_k x^k, \qquad (g_0 > 0)$

We have,

$$\boxed{\left|F(0)e^x - F(x)\right| < e^{(x)}\left|f(x)\right|}$$

Put $\mathbf{Q}(x) = F(0)\, e^x - F(x)$, that

$$\sum_{k=0}^{n} g_k \theta(k) = F(0)\sum_{k=0}^{m} g_k e^k - \sum_{k=0}^{m} g_k F(k)$$

Using to (2), this yields

$$\sum_{k=0}^{m} g_k \mathbf{Q}(k) = -\sum g_k F(k)$$

In order to prove the transcendence of $e$ we choose the auxiliary function $f(x)$ as in the above lemmas. Suppose, on the contrary, that $e$ was not transcendental. Then there exists a polynomial with $e$ as a root.

$$G(x) = \sum_{\nu=0}^{m} g_\nu x^\nu \qquad G(e) = \sum_{\nu=0}^{m} g_\nu e^\nu = 0 \quad (g > 0)$$

Put $\theta(x) = F(0)\, e^x - F(x)$, from which we get

$Q(k) - F(0)\, e^k - F(k)$, $(k = 1, ..., p)$. Hence

$$\sum_{k=0}^{m} g_k Q(k) = F(0)\sum_{k=0}^{m} g_k e^k - \sum_{k=0}^{m} g_k F(k)$$

i.e. $$\frac{1}{(p-1)!}\left|\sum_{k=0}^{m} g_k Q(k)\right| = \left|\sum_{k=0}^{m} g_k F(k)\right|\frac{1}{(p-1)!}$$

We restrict the prime $p$ further by the condition $p > g_0$ on the right hand side of the above equation, the term for $k = 0$ is

$g_0 \dfrac{1}{(p-1)} F(0) \not\equiv 0 \pmod p$. Consequently the left side is an *integer* $\neq 0$. Thus

$$\frac{1}{(p-1)!}\left|\sum_{k=0}^{\nu} g_k F(k)\right| \geq 1.$$

(4) $$\left|\frac{1}{(p-1)!}\sum_{k=0}^{m} g_k Q(k)\right| = \left|\frac{1}{(p-1)!}\sum_{k=0}^{m} g_k F(k)\right|$$

Herein, we have

$\frac{1}{(p-1)!}F(0) \not\equiv 0 \pmod p$ and $\frac{1}{(p-1)!}F(k) \equiv 0 \pmod p$, ( $k$ = 1, ...$n$) so that the sum on the right side of (4) is a *positive* integer, thus

(5) $$\frac{1}{(p-1)!}\left|\sum_{k=0}^{m} g_k F(k)\right| \geqq 1$$

We show that the left side of (4) is < 1. This absurdity will show that our initial hypothesis is wrong. It remains, therefore, to prove the following lemma, where we prove a stronger statement.

**Lemma**

(6) $$\lim_{p\to\infty}\frac{1}{(p-1)!}\sum_{k=0}^{m} g_k Q(k) = 0$$

**Proof of the Lemma:** We have the estimate

$$|Q(k)| \leqq e^k |f(k)|.$$

Now $|f(k)| \leqq k^{p-1}(k)^p \ldots (k+n)^p$. Put

$$u_p = \frac{k^{p-1}(k+1)^p \ldots (k+n)^p}{(p-1)!}.$$ Then the

$\frac{u_{p+1}}{u_p}\frac{k_{(k+1)}\ldots(k+n)}{?} \to 0$ as $p \to \infty$. By D'Alembert's Test, this gives $u_p \to 0$ as $p \to \infty$. We have in (6) only a *finite* number of sequences, each of which tends to zero. Hence follows the Lemma.

By taking $p$ large, we $u_c$ sum on the left side of (6) can be made as small as we please. It is quite enough for us to take the *prime p* so large that it is, in absolute value < 1.

In order to prove the weaker statements that $e$ and $\pi$ are *irrational* there are simpler proofs.

**Theorem:** $e$ is irrational.

**First proof:** We have $e^x = 1 + \frac{x}{1!} + \frac{x^2}{2!} + ...$ whence

$$e^x = 1 - \frac{1}{1!} + \frac{1}{2!} + .. \text{ we can write}$$

$$\frac{1}{e} = e^{-1}. \text{ Put } s_n = \sum_{k=0}^{n} \frac{(-1)^k}{k!}, \text{ and } r_n = \sum_{k=n+1}^{\infty} \frac{(-1)^k}{k!}$$

$$0 < (-1)^{n+1} r_n = \frac{1}{(n+1)!} - \frac{1}{(n+2)!} + ...$$

$$= \frac{1}{(n+1)!} - \left(\frac{1}{(n+2)!} - \frac{1}{(n+3)!}\right)$$

$$< \frac{1}{(n+1)!}$$

Therefore $0 < n!r_n(-1)^{n+1} < \frac{1}{n+1} < 1$ so that $n!e^{-1} = n!\sigma_n + n!$ $r_n(-1)^{n+1}$ can never be an integer. So $e^{-1}$ cannot be rational.

**Second proof:** We have

$e = \sum_{k=0}^{\infty} \frac{1}{k!} = s_n + r_n$ where $s_n = \sum_{k=0}^{n} \frac{1}{k!}$ and $r_n = \sum_{k=n+1}^{\infty} \frac{1}{k!}$. Here, putting

$A_n = = n!r_n$ and $\vartheta_n = n!r_n$ we see that $A_n$ is an *integer*, and

$$r_n = n!r_n = \frac{1}{n+1} + \frac{1}{(n+1)(n+2)} + ... < \frac{1}{2} + \frac{1}{2^2} + ... = 1$$

Hence $x = A_n + r_n$, $(0 < \vartheta_n < 1)$. Therefore $[x] = A_n = n!r_n < n!e = x$, i.e. $[x] < x$. If $e$ were rational, we can always find $n$ such that $x = n!e$ is an integer which contradicts $[x] < x$. ■

**Theorem** $\pi$ is irrational.

**Proof:** Assume, on the contrary, that $\pi$ is rational: $\pi = \frac{a}{b}$, $a > 0$, $b > 1$ integers. We consider the auxiliary function

(1) $f(x)=\dfrac{x^n}{n!}(a-bx)^n$

and set $F(x)=\sum_{k=0}^{2n}(-1)^k f^{(2k)}(x)$; thus we have

(2) $F(x)=f(x)+-...-+(-1)^{n-1}f^{2n-2}(x)\,(-1)^{(n)}f^{(2n-2)}(x)+(-1)^n f^{(2n)}(x)$

Differentiating twice, and noticing $f^{(2n+2)}(x)\equiv 0$, we have

$F''(x)=t''(x)=f''(x)+-...+(-1)^{2n-2}f^{2n}(x)+0$

whence we have

(3) $F(x)+F''(x)=f(x)$

and $\dfrac{d}{dx}[F'(x)\sin x-F(x)\cos x]$

$$=\big(F'(x)+F(a)\big)\sin x=f(x)\sin x$$

Consequently

$$\int_0^{\pi} f(x)\sin x\,dx=[F'(x)\sin x-F(0)\cos x]_0^{\pi}$$

i.e. $\displaystyle\int_0^{\pi} f(x)\sin x\,dx=F(\pi)+F(0)$

since $f(x)=\dfrac{x^n}{n!}(a-bx)^n=\dfrac{b^n x^n}{n!}(\pi-r)^n$, (recall our assumption that $\pi=\dfrac{a}{b}$.)

Hence $f(x)=f(\pi-x), f'(x)=-f'(\pi-x), f''(x)=f''(\pi-x)$ so that

$$F(x)=\sum_{k=0}^{2n}(-1)^k\,\delta^{2k}(x)=F(\pi-x)$$

whence we obtain that $F(x)=F(0)$. Hence (4) becomes

(5) $\boxed{\displaystyle\int_0^{\pi} f(x)\sin x\,dx=2F(0)}$,

$$f(x)=\left(f_n(x)=\frac{x^n}{n!}(a-bx)^n\right)$$

Let $x_0$ be an interior point of the interval $[0, \pi]$. Since $f(x)$ is continuous at $x_0$, for every $\varepsilon > 0$, there exists a neighbourhood. $J$ of $x_0$ which is $\subset [0, \pi]$ such that

$$|f(x) - f(x_0)| < \varepsilon \quad \text{for } x \,\varepsilon\, J$$

Let us choose $\varepsilon = \dfrac{f(x_0)}{2}$; then

$$f(x_0) - \varepsilon < f(x) < f(x_0) + \varepsilon,$$

and therefore $\quad f(x) > f(x_0) > 0$ for $x \,\varepsilon\, J$.

Accordingly

$$\int_0^\pi g(x)\,dx > \int_J g(x)\,dx > 0$$

which proves the Lemma.

Looking back at (5) we show that the left side $\to 0$ as $n \to \infty$, and hence can be made $< 1$ by choosing $n$ sufficiently large. Since the right side of (5) is a positive integer this brings the contradiction and completes the proof of the theorem.

We have that $f(x) = \dfrac{x^n}{n!}(a - bx)^n$; using Taylor expansion on the left side, and the binomial expansion on the right side we obtain that

$$\sum_{k=0}^{2n} \frac{f^{(k)}(0)}{k!} x^k = \frac{x^n}{n!} \sum_{\nu=0}^{n} (-1)^\nu \binom{n}{\nu} a^{n-\nu} \cdot b^\nu x^\nu$$

Comparing like powers of $x$, we see

$$\frac{f^{(k)}(0)}{k!} = 0 \text{ for } 0 \le k \le n \text{ and } f^{(n+\nu)}(0)$$

$$= (-r)^\nu \binom{n}{\nu} a^{n-\nu}\, b^{n-\nu} \cdot (n+\nu)!$$

$$(\nu = 0, 1, ..., n)$$

Consequently

(6) $F(0)=\sum_{k=0}^{n}(-1)^k f^{(2k)}(0)$ is an INTEGER

We need that F(0) to be a *positive* integer. Comparing (5) and (6) it is enough to prove that

**Lemma: Given:** 1° g ($x$) is **continuous** in the closed interval $0 \le x \le \pi$....; 2° $g(x) > 0$ in the open interval $0 < x < \pi$.

**Claim**

$$\int_0^\pi g(x)\,dx > 0$$

Let $x_0$ be an interior point of the interval $[0, \pi]$. Since $f(x)$ is continuous at $x_0$, for every $\varepsilon > 0$, there exists a neighbourhood. $J$ of $x_0$ which is $\subset [0, \pi]$ such that

$$\left|f(x)-f(x_0)\right| < \varepsilon \qquad \text{for } x \in J$$

Let us choose $\varepsilon = \dfrac{f(x_0)}{2}$; then

$$f(x_0)-\varepsilon < f(x) < f(x_0)+\varepsilon,$$

and therefore $f(x) > f(x_0) > 0$ for $x \in J$.

Accordingly

$$\int_0^\pi g(x)\,dx > \int_J g(x)\,dx > 0$$

which proves the Lemma.

Looking back of (5) we show that the left side $\to 0$ as $n \to \infty$, and hence can be made $< 1$ by choosing n sufficiently large. Since the $p$ right side of (5) is a positive integer this brings the contradiction and completes the proof of the theorem.

Put $u_n = f(x)\sin x = \dfrac{x^n}{n!}(a-bx)^n \cdot \sin x$.

Then $\dfrac{u_{n+1}}{u_n} = x(a-bx)\cdot\dfrac{1}{n+1} \to 0$ as $n \to \infty$. By D'Alambert's test we get, therefore, $u_n \to 0$ as $n \to \infty$. Hence 0, $u_n < \frac{1}{\pi}$ for $n > N_0$.

Accordingly

$$\int_0^{\pi} f(x) \sin x \, dx < \frac{1}{\pi} \int_0^{\pi} dx = 1$$

The proof of the theorem is complete. ■

## § 6. DIRICHLET'S APPROXIMATION THEOREM. SIEGEL'S LEMMA

If $\vartheta$ is any irrational number, the simplest rational approximation to $\vartheta$ is obtained as follows. We take any natural number $N$, we choose $p = [\vartheta N]$ and $q = N$. Then

(1) $$\boxed{\left|\vartheta - \frac{p}{q}\right| < \frac{1}{N}}$$

Dirichlet, by a special device, sharpened this inequality: Divide the interval [0,1] into subintervals

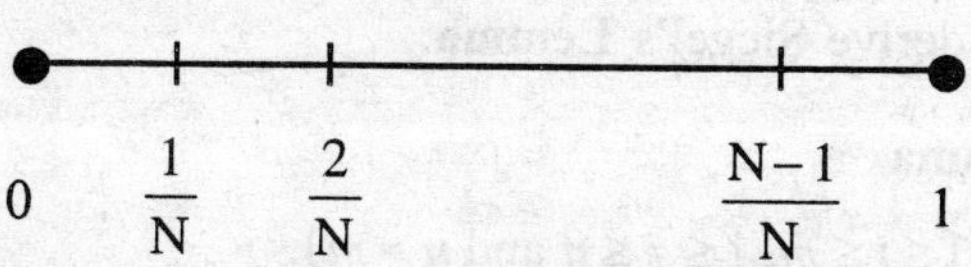


$J_k : \frac{k-1}{N} \le x < \frac{k}{N} (k = 1, ..., N)$. Then, the fractional parts $\{n\vartheta\}$ $= n\vartheta$ [ $[n\vartheta]$, ($n$ = 0, 1, ..., N) must all be in [0,1]. These ($N$+1) fractional parts fall on these subintervals $J_k$, ($k$ = 1, ..., $N$). Hence there must exist $\{a\alpha\}$ and $\{b\alpha\}$ with $0 \le a < b \le N$ falling in one and the same $J_k$:

$$\frac{k-1}{k} \le \{a\alpha\} < \frac{k}{N} \quad \text{and} \quad \frac{k-1}{N} \le \{b\alpha\} < \frac{k}{N}$$

$$\frac{k-1}{k} - \frac{k}{N} < \{b\alpha\} - \{a\alpha\} < \frac{k}{N} - \frac{k-1}{N}$$

$$-\frac{1}{N} < (b\alpha - [b\alpha]) - (b\alpha - [a\alpha]) < \frac{1}{N}$$

Let $q = (b - a)$, and $p = [b\alpha] - \lfloor a\alpha]$, then

$$-\frac{1}{N} < q\alpha - p < \frac{1}{N}$$

(2) i.e. $$\boxed{\left|\alpha - \frac{p}{q}\right| < \frac{1}{qN}}$$

It is obvious that (2) is a sharpening of (1).

**Remark:** We see from (20 that there exist infinitely many rational numbers $\frac{p}{q}$ such that $\left|\alpha - \frac{p}{q}\right| < \frac{1}{q^2}$.

The method that was used in proving (2) is often used in Diophantine Approximation. It is known as Dirichlet's "Schubfach Prinzip" or "Box Principle". If there is a set **A** of $n$ objects, and $m$ boxes with $n > m$, and if these objects have to be put in those boxes, then there exists at least one box containing at least two elements. Speaking in the language of set theory, if **A** is a set of $n$ elements, and **B** is a set of $m$ elements with $n > m$ no map $\varphi : A \to B$ cannot be one-to-one. The "Box Principle", obvious as it sounds, does many things in Diophantine Approximation. As a further application of this principle we derive Siegel's **Lemma**.

**Siegel's Lemma**

Let $a_{ij} \in \mathbf{Z}$, $(1 \le i \le m, 1 \le j \le n \text{ and } n > m)$.

$M = \max_{i,j} |a_{i,j}|$. Then there exists $x = (x_1, ..., x_n) \in Z^n$ such that

(3) $y_{ij} = a_{i1}x_1 + ... + a_{in}x_n = 0, (i = 1, ..., n)$ and

(4) $\boxed{|x_i| < (nM)^{\frac{n}{n-m}}}$

**Preliminary Remark.** If $a_{ij} \in Q$, the linear homogeneous system (3) of equations has always a solution $x = (x_1, ..., x_n) \neq (0, ..., 0)$, because the number of unknown is greater than the number of equations $(n > m)$. After a rational solution $\left(\frac{a_1}{b_1}, ..., \frac{a_n}{b_n}\right)$ is obtained, by multiplying by the l.c.m. of $b_1, ..., b_n$ we can get a solution in $\mathbf{Z}^n$. The point of the lemma is in obtaining an estimate (4).

**Proof of Siegel's Lemma**

Denoting by $[\xi]$ the integral part of a real number $\xi$, put $C = [(nM)^{m/n-m}]$ so that $(C+1)^{n-m} > (nM)^m$, i.e.

(5) $(C+1)^n > (nm\,(C+1)^m)$

Let us call a point $x = (x_1, ..., x_n) \in Z^a$ a **lattice point**. Denote by $E$ the set of lattice points with $0 \le u \le C$ $(n = 1, ..., n)$. Since there are $C + 1$ choice for each $q_i$, $(i = 1, 2, ..., n)$, we have

(6) $\#E = (C+1)^n$

(where # E denotes the number of elements in $E$). We wish to find $\#\,\varphi\,(E)$ where $\varphi$ is the map $\varphi : Z^n \to Z^m$ is given by $\varphi\,(x_1, ..., x_n) = (y_1, ..., y_n)$. We have $y_j = q_{ji}x_1 + ...\, a_{jn}x_n$. Let

$$\varepsilon_k = \begin{cases} +1 \text{ if } a_{jk} > 0 \\ -1 \text{ if } a_{jk} < 0 \end{cases}, \; y_j' = \sum_{k=1}^{n} \varepsilon_k a_{bk} x_k .$$

Here each term in $y_j'$ is > 0, and $\varepsilon_k\, a_{jk} = |a_{jk}| \le M$ and

$0 \le x_k \le C.$ i.e. there are $nMCH$ points $y_j'$, hence also $(nMC + 1)$ points $y_j$. Accordingly, (Sec (5)).

(7) $\#\,E > \#\,\varphi\,(E)$.

By Dirichlet's principle, there must exist two **distinct** $x'$ and $x''$ such that $\varphi\,(x') = y = \varphi\,(x'')$. If we put $x = x' - x''$, we have $\varphi\,(x) = 0$ $(x \ne 0$ in $Z^n)$.

Since $0 \le x_j' \le C, 0 \le x_j'' \le C, |x_j| = |x_j - x_j| \le C.$

The Lemma is proved. ■

## § 7. MISCELLANEOUS THEOREM AND EXAMPLES

I. For $\alpha \ne 0$, the numbers $\alpha$ and $e^{\alpha}$ cannot both lie in the Gaussian field $\theta$ (i), $i = \sqrt{-1}$. Taking $\alpha = 1$, we see that $e \notin 0(i)$, *i.e.*, $e \notin \theta$. $e$ is irrational. Since $e^{\pi i} = -1$, we see $\pi_i \notin \theta$, $\pi \notin \theta$.

II. For $\alpha, \beta \ne 0, 1$, $\alpha$ algebraic and $\beta \in \underline{Q\,(i)}$, $\alpha^p$ is transcendental.

Thus $(-1)^{-i} = e^{-i\log(-1)} = e^{-i(i\pi)} = e^{\pi}$ is transcendental.

III. For $\alpha$, $\beta$ algebraic $\neq 0, 1$, $\alpha^{\beta}$ is transcendental (Theorem of Gelfond/Schneider). This gives the solution of the **seventh problem** of Hilbert.

IV. The Riemann-Zeta function $\zeta(s) = \sum_{n=1}^{\infty} \frac{1}{n^s}$ converges for Re $s > 1$. We have, for $k = 1, 2, 3, \ldots$

$$\boxed{\zeta(2k) = (-1)^{k+1} \frac{1}{2(2k)!} B_k}$$

where the $B_k$ are certain *rational* numbers called *Bernoulli Numbers.*

The Bernoulli Numbers appear as coefficients in the Taylor expansion of the function $z/e^Z - 1$. For example

(i) $B_0 = 1, B_1 = -\frac{1}{2}, B_2 = -\frac{1}{6}, B_4 = -\frac{1}{90}, \ldots$

(ii) $B_n = 0$ for $n$ **odd** $\geq 3$.

(iii) $\zeta(2) = \pi^2/6$.

The arithmetic nature of $\zeta(2k+1)$, $k = 1, 2, \ldots$ was not known. Recently, Apèry proved that $\zeta$ (3) is *irrational*. See the following references.

(1) **J. Vander Poorten.** A proof that Euler missed ... Apery' of the irrationality of $\zeta$ (3). Resonance, 1/March/2003

(2) **F. Beukers.** A note on the irrationality of $\zeta$ (2) and $\zeta$ (3). Bull. London Math. Soc. **11** (1979), pp. 268-272.

(3) **S. Chowla.** Congruence properties of the Apèry Number. Journal of Number Theory 12, pp.188-190 (1980).

(Mahler's Rational Approximation to $\pi$)

1° For all rational numbers $p/q$, there holds the inequality

$$\boxed{\left|\pi - \frac{p}{q}\right| \geq q^{-42}} \quad \text{(Mahler)}$$

2° If $\alpha$ is an algebraic number # 0,1,

$$\boxed{|\log \alpha - \beta| > B^{-c}} \qquad \text{(Baker)}$$

where $\beta$ runs through all algebraic numbers of degree $\leq d$, and height $\leq$ B (B $\geq$ 2). In particular $|\pi - \beta| > B^{-C}$, (C depends only on deg $\beta$).

$$\boxed{\left|e^{\pi} - \frac{p}{q}\right| > q^{-c} \log\log q} \quad \text{(Baker)}$$

**Reference: A Baker** : Transcendental Number Theory, Cambridge University Press (1975), Ch 3, p.38.

**Thue Approximation Theorem**

Let $\vartheta$ be an algebraic number of degree $\geq 3$. Let $\varepsilon > 0$, $C > 0$. Then there exist only *finitely many* pairs of integers $(p, q)$, $q > 0$, satisfying the inequality

$$\boxed{\left|\vartheta - \frac{p}{q}\right| < \frac{c}{q^{n/2+1+\varepsilon}}}$$

**Special case:** ($d = 3$, $\varepsilon = 1/2$)

**Theorem:** Let $D$ be an integer which is not a perfect cube, and $\vartheta = \sqrt[3]{D}$. Then there exist only *finitely many* pairs $p$ and $q$ of integers satisfying the inequality

$$\boxed{\left|\vartheta - \frac{p}{q}\right| < \frac{C}{q^3}}$$

we recall, while obtaining Liouville's theorem we used the minimal defining polynomial $f^*(x)$ of $\vartheta_1$ ($d = \deg \vartheta$). We needed for the proof the two crucial properties of $f^*(x)$ namely

$f^*(x)$ vanished at $\vartheta$, and did not vanish at $(p/q)$ :

$$f^*(\vartheta) = 0,\ f^*\left(\frac{p}{q}\right) \neq 0.$$

For the proof of Thue's theorem we need a polynomial of two variables which must satisfy the conditions: Let

$$F(x, y) = \sum_{i=0}^{m+n} \sum_{j=0}^{s} a^{ij} x^i y^j \qquad F^{(\lambda)}(x, y) \frac{1}{\lambda!} \frac{\partial^\lambda}{\partial^x} F(x, y)$$

Then the conditions are

(1) $F^{(x)}(\vartheta, \vartheta) = 0$ for $r = a, 1, 1, \ldots, r-1$

In other words $F(x, \vartheta) = (x - \vartheta)^r G(x)$

Assuming, contrary to the theorem, that there exist infinitely many rational numbers $p_n/q_y$ $(n = 0, 1, 2, \ldots)$ we want to show

$$(2) \quad \boxed{F\left(\frac{p_1}{q_1}, \frac{p_2}{q_2}\right) \neq 0}$$

The conditions imposed on the polynomial $F(x, y)$ yield a system of linear homogeneous equations for its coefficients. Siegel's Lemma steps in here to our help.

To show that $F\left(\frac{p_1}{q_1}, \frac{p_2}{q_2}\right) \neq 0$ is the hardest part of the proof.

From Thue's Approximation Theorem, we can deduce the following:

**Theorem**

Let D be a positive integer which is not a perfect cube. Then the equation

$$\boxed{x^3 - Dy^3 = 1}$$

has only *finitely* many solutions in integers $x$ and $y$.

**Remark**

We have seen that the Bhaskara equation $x^2 - Dy^2 = 1$, ($D > 0$ *not* a perfect square) has *infinitely* many solutions. There is a fundamental solution $(x_0, y_0)$ which give all the solutions $(x_n, y_n)$ are obtained by

$$x_n + \sqrt{D} y_n = \left(x_0 + \sqrt{D} y_0\right)^n.$$

The simple continued fraction expansion of $\sqrt{D}$ gives these results. Even in the case of the equation $x^3 - 2y^3 = 1$, we are not able to find the finitely many integer solutions by means of the simple continued fraction of $\sqrt[3]{2}$ because there is no effective algorithm for finding the expansion, as in the case of $\sqrt{2}$.

**Remarks**

We would like to suggest to the student reader to study the inspiring presentation of the special case of Thue's Theorem given in

J.H. Silverman, J. Tate : Rational Points on Elliptic Curves, Springer Verlag (1992).

## REFERENCES

1. Sri Bhārati Krishna Tirthaji Maharaja: Vedic Mathematics, Motilal Banarsi Das, new Delhi (1978).
2. Sri Bhārati Krishna Tirthaji Maharaja: Vedic Metaphysics, Motilal Banarsi Das, new Delhi (1978).
3. B.B. Datta and A.N. Singh: History of Hindu Mathematics, Parts I and II, Motilal Banarsi Das, (1935), (1938).
4. Krishna Ganaka Bija Pallavam: a commentary on Bija Ganita of Bhaskara.
5. T.A. Sarasvati Amma: Geometry in Ancient and Medieval India, Motilal Banarsi Das, (1979).
6. C.N. Srinvasiengar: History of Ancient Indian Mathematics. The World Press Private Ltd., Calcutta, (1967).

### Suggested Further Reading

1. Alan Baker: A concise introduction to the theory of numbers, Cambridge University Press, England (1984).
2. Loo Keng Hua: Introduction to Number theory, Springer-Verlag Berlin (1982).
3. I.M. Vinogradov: Elements of Number theory, New York, Dover (1954).
4. S. Chowla: The Riemann Hypothesis and Hilbert's Tenth Problem, Chapter 8, pp. 81–105 Gordon and Breach, New York (1965).
5. L.J. Mordell: Diophantine equations, Academic Press, New York (1969).
6. E. Hlawka: Theories der Gleichverteibung B.I. Wissenschafts vertag, Mannheim/Wein/Zurich, (1979).
7. L.J. Mordell: On a Pellian Equation conjecture (I) *Acta Arithmetica* **6** (1960) 137–144.
8. L.J. Mordell: On a Pellian Equation conjecture (II) *J. London Math. Soc* **36** (1960), 282–288.
9. L.J. Mordell: On the four integer cube problem *J. London Math. Soc* **II** (1936), pp. 208–218.

*Remark*

The papers 7,8,9 of L.J. Mordell are related to the following conjectures regarding the Bhāskara equation

$Dx^2 - 1 = y^2$, $D > 0$, $D \neq$ square of an integer.

Conjecture I [Ankeny-Artin-Chowla]

Let $D$ be a prime $\equiv 1(mod 4)$, and $(x_0, y_0)$ the solution of the equation, $Dx^2 - y^2 = 4$ with the least positive value of $x$, then $x_0 \not\equiv 0 \left(mod D\right)$.

Conjecture II

Let $D$ be a prime $\equiv 3(mod 4)$, and $(x_0, y_0)$ the solution of the equation, $Dx^2 - y^2 = -1$ with the least positive value of $x$, then $x_0 \not\equiv 0 \left(mod D\right)$.